THE BIOLOGY TEACHER'S
SURVIVAL GUIDE

For my brother, Dave,
and his wife, Susan

THE BIOLOGY TEACHER'S
SURVIVAL GUIDE

Michael F. Fleming

Order this book online at www.trafford.com
or email orders@trafford.com

Most Trafford titles are also available at major online book retailers.

Print information available on the last page.

ISBN: 978-1-4907-5770-4 (sc)
ISBN: 978-1-4907-5769-8 (e)

Trafford rev. 03/31/2015

North America & international
toll-free: 1 888 232 4444 (USA & Canada)
fax: 812 355 4082

About This Resource

Biology Teacher's Survival Guide was written for the sole purpose of making your teaching of biology fun, enjoyable, and profitable for you and your students. It is packed with scores of novel and innovative ideas and activities that you can put to use immediately in your classroom.

Section 1, "Innovative Classroom Techniques for the Teacher," presents a variety of techniques to help you stimulate active student participation in the learning process. It includes, among other things, an alternative to the written exam, a technique to elicit student responses to questions and discussion topics, ideas to develop students' appreciation of career possibilities, a technique to keep students' attention focused on the task at hand, and a way to allow students to participate in the correction of their tests.

Section 2, "Success-Directed Learning in the Classroom," explains how you can easily make your students accountable for their own learning. Techniques presented focus on teamwork between you, the student, and the parent or guardian in assuring student achievement and success; the goal is no marking period grade below a C. Individual achievement is placed in the student's hands, and your role of "villain" in the grading process is virtually eliminated.

Section 3, "General Classroom Management Techniques," addresses and provides solutions to a wide variety of classroom management issues that you face in creating an ideal rapport with students. Such issues include the student opposed to dissection, student lateness to class, and the chronic discipline problem. This section also explores such topics as self-evaluation techniques, keeping current in subject matter content, preventing teacher burnout, and the art and science of teaching.

Section 4, "An Inquiry Approach to Experiencing the Spirit and Nature of Science," details an effective approach to teaching scientific inquiry (the spirit) and research technique (the nature). It is complete with three fully planned inquiry units that you can begin using with the basic materials of test tubes, graduated cylinders, yeast, and molasses. Please be sure to read the list of important safety caution concerns in 4.2 first.

Section 5, "Sponge Activities," gives you 100 *reproducible* activities you can use at the beginning of, during, or at the end of class periods. These activities cover a wide range of biology topics including the cell, classification, plants, animals, protists, the microscope, systems of the body, anatomy, physiology, genetics, and health. Moreover, the format of the activities ranges from crosswords and matching to message squares and fill in the blanks.

To help you quickly locate appropriate worksheet activities in Section 5, the Contents includes a list of all 100 worksheets in this section in alphabetical order, from Algae (Worksheet 5-1) through Vitamins and Minerals (Worksheet 5-100).

This entire book is supplemented with reproducible teacher and student worksheets and evaluation forms, all designed to streamline your classroom management and free your time for accomplishing the numerous other tasks that you face. You can photocopy any of these reproducibles as many times as required for use with individual students or an entire class.

For the beginning teacher who is new to the classroom as well as for the more experienced teacher who may want a "new lease on teaching," this resource is designed to bring fun, enjoyment, and profit into the teacher–student rapport that is called teaching.

Michael F. Fleming

About the Author

MICHAEL F. FLEMING, ED.D., The Pennsylvania State University, has taught general science, basic biology, anatomy and physiology, microbiology, and behavioral science for over 30 years in the Council Rock School System, Newtown, Pennsylvania. He has participated in curriculum development and is vitally interested in new and novel approaches to motivating students to experience learning as an ongoing and enjoyable process.

Dr. Fleming has been awarded National Science Foundation summer grants and research grants from the Heart Association of Southeastern Pennsylvania. He has also presented papers at conventions of the National Association of Biology Teachers.

In addition, Dr. Fleming created and has carried out for over 15 years a voluntary student participation activity, known as The Anatomy Awards, involving the Leukemia Society of America, Inc. This activity engages students in designing costumes based on structures studied in biology and wearing the costumes in community-related activities to raise money for the Society.

Dr. Fleming has published several articles in professional journals, including "Protists, Photomicrographs and Sterezooms: A Study Unit in Biology" in *Focus*, and "The S-A-S (Science As Science) Approach to Teaching Second Level Biology," "Organ Alley: Pathway of Learning in a Biology Classroom," and "The Incredible Edible Model: Food for Thought," in *The American Biology Teacher*. He is also the editor and coauthor of the "Science Project Cards" series, the *Life Science Labs Kit* (1985), and *Science Teacher's Instant Labs Kit* (1991), all published by the Center for Applied Research in Education.

Michael F. Fleming, nephew of the author and artist for this activity book, is a graduate of the John Herron School of Art at I.U.P.U.I. in Indianapolis, Indiana.

Acknowledgments

Sincere thanks and appreciation to Sandra Hutchison and Connie Kallback, whose generous editorial help was always available and invaluable. Thanks also to Barbara O'Brien, whose assistance was invaluable in the fine-tuning of the manuscript. Special thanks to my nephew, Michael F. Fleming, whose artwork is a major contribution to this book.

Contents

SECTION 2: SUCCESS-DIRECTED LEARNING IN THE CLASSROOM 107

SECTION 5: Sponge Activities 203

M

N

P

R

S

T

U

V

SECTION 1

Innovative Classroom Techniques for the Teacher

The first section of the Biology Teacher's Survival Guide presents 24 innovative techniques you can begin to use at once to motivate real student interest in the learning process. These are accompanied by over 50 reproducible worksheets to help you implement the suggested techniques.

Among other things, you'll find an alternative to the written exam (voluntary oral exams, or VOEs), a way to elicit student responses to questions and discussion (Pic-a-Tag Technique), and a method to allow students to participate in the correction of their tests (instant corrections, or ICs).

1.1 THE MESSAGE SQUARE ACTIVITY _____

Perhaps the best way to introduce your students to the Message Square activities that appear in Section 5 of this resource is to select one of the activities, prepare an overhead transparency of it, and project it on the classroom screen. Next, review the following procedure to be used in solving each Message Square Activity. As you explain each step of the procedure, illustrate using the overhead transparency.

1. Using a 3 × 5 card, cover the instructions that direct you to write particular letters in particular squares.
2. Read the hint.
3. Move the 3 × 5 card, revealing one letter at a time, and write that letter in the proper square(s).
4. At any step along the way that you think you know the answer, write it in the anticipated-answer column before proceeding. (This step is important, because the temptation for the student is to fill in the squares quickly, disregarding this step.)

The Message Square Activity can be used

- as an individual student activity, or
- as a full-class exercise using the overhead projector.

The Research Questions for Text or Library can also be used in various ways:

- as an in-library activity for helping the entire class practice library skills
- as an extra-credit reward for the first student to answer the research questions correctly (of course, the student must provide the source of the answer). You might continue to award extra credit to other students, as long as different sources are used in obtaining the correct answers.

1.2 THE PIC-A-TAG TECHNIQUE _____

At times a facilitator is needed to catalyze student participation in the classroom. One effective facilitator that students enjoy is the Pic-a-Tag Technique. It keeps students alert because anyone's number can come up, and it accomplishes this in the spirit of classroom cooperation, not coercion.

The Pic-a-Tag Technique allows you to select students randomly (eliminating student complaints of "you always call on me") for such activities as answering questions, going to the chalkboard, participating in a lab demonstration, and oral reading of classroom articles.

Key tags are excellent to use for the Pic-a-Tag Technique. For example, if you have 25 students in your class, simply number the key tags from 1 to 25. Place the tags in a small container in which they can be mixed.

The following example of the Pic-a-Tag Technique assumes that students are sitting at desks or lab tables arranged in a rectangle. If your students are sitting in a circle or rows or some other arrangement, the counting procedure can be easily modified. Let's assume that you need a student to answer a question, and you decide to use the Pic-a-Tag Technique. You announce to the class that you are going to pick a tag and start counting from, for example, the back left of the classroom. If the number picked is even, you might want to count in a clockwise direction; if odd, in a counterclockwise direction. If you then need another student, pick another tag and start the count from the previous student selected.

There are several variations that you can use with the Pic-a-Tag Technique. For example, you might wish to ask for a volunteer prior to picking a tag. If a student volunteers, then he or she can select the next student needed. Another variation is to allow any student selected by a tag to choose the next student needed.

Students enjoy the technique of being randomly selected by tags and having the opportunity to personally select other students. Indeed, when given a choice, whether to use the Pic-a-Tag Technique or not, students will invariably select Pic-a-Tag.

1.3   MUSIC, FOOD, AND BEVERAGE IN THE CLASSROOM _____

Having a radio or tape player in the classroom can do a lot to enhance student enjoyment of learning. Music can provide a relaxing ambiance in which to carry out lab activities or accomplish desk work. The volume can easily be kept at an acceptable classroom level so that it does not interfere with occasional teacher instructions.

You might wish to expose your students to some alternative music types, such as classical or light jazz, from time to time. Other than that, students can be allowed to select the music that they desire. It can also be fun to give each student an opportunity to select the day's music.

It would be a rare administration that would not allow music in the classroom. However, when it comes to food and beverage (other than edible models), it would be a rare administration that would approve.

It can be argued that there are times when permitting some food and beverage can be advantageous to the learning situation. It is yet another way to make learning more enjoyable for students. Having a soft drink or a cup of tea or hot chocolate during desk work does nothing to interfere with students' learning. Enjoying a sandwich or a candy bar during desk work is not disastrous.

SAFETY NOTE

Obviously there are class activities, such as those involving potentially dangerous laboratory materials, during which it would not be safe to allow food or beverage under any conditions.

As long as students clean up after themselves, what can be the harm of a classroom snack during suitable class activities?

It is important for you to strive continually to make the classroom ever more pleasant and enjoyable for students during their daily 40- or 50-minute stay.

1.4   VOEs (VOLUNTARY ORAL EXAMS)

Voluntary oral exams are, as implied, voluntary on the part of students. You might want to administer them one or two days prior to when you plan to give the entire class a written test.

The VOE provides a refreshing avenue for student evaluation, and once students become used to the idea of oral exams, many will take advantage of them.

VOE questions are prepared on 3 × 5 cards (one question per card) and cover all materials on which students are to be tested. Questions often range from recall of textbook information to identification of structures on actual laboratory models or on diagrams projected on the classroom screen. The prepared cards can, of course, be used from year to year.

The student volunteer comes to the front of the classroom. If you like to "ham it up," consider having a lectern, microphone, and microscope-lamp "spotlights." Students enjoy this bit of levity.

Beginning with the top 3 × 5 card, read the question to the student. If the student answers five out of six questions accurately, he or she earns an exam grade of A. There is no penalty if the student does not correctly answer five questions (the student earns an A or nothing).

As incentives for students to study and volunteer for VOEs, consider the following:

- A student who earns an A on the VOE is exempted from taking the written exam. The student does, however, have the option of trying for another A on the written exam. If the student chooses that option, he or she must take the grade received on the written exam even if it is lower than an A.
- A student who earns an A on 80% of the VOEs given during the year (for example, 8 out of 10) is exempted from the final.

In whatever way you wish to incorporate VOEs in your program, you will find them an excellent motivational technique.

1.5   ICs (INSTANT CORRECTIONS OF TESTS) _____

Students usually hope that tests will be corrected and returned as soon as possible. Your goal might be to return tests the day after they are taken, but often this is not feasible. However, there is a technique that will (a) provide some students with immediate test grade feedback, and (b) allow you to get many of the tests corrected during the same period they were given.

Plan tests that allow for some time left over at the end of the class period. During this remaining time, students can request an IC (instant correction) of their tests. You will get plenty of requests, and students will appreciate the opportunity for an IC.

You might want to allow each student to watch as you correct his or her test. This creates a rapport between you and the student and allows you to point out such common features as unclear answers, poor word choices, and poor legibility. You can also question the student in more depth, if necessary, when evaluating an answer. Overall, an IC is a rather fast process, and you can get many tests corrected in a short period of time.

1.6   THE EDIBLE MODEL _____

One of the major goals of teachers is to make information as interesting and stimulating as possible. Teachers are always seeking innovative techniques to make learning genuinely enjoyable. The edible model will help achieve these goals. Edible models will rapidly become your students' favorite extra-credit activity, and students will create some truly ingenious models.

The edible model is a model, constructed by a student, that depicts something studied in class. It could be, for example, a particular body structure such as a stomach, heart, or brain. It might be of a body tissue such as muscle, bone, or nerve. It also might be of a piece of laboratory equipment such as a microscope.

The edible model is prepared using any edible materials. Materials can range from cakes, pies, and pizzas to pancakes and licorice sticks. The materials should be assembled in such a manner that the resulting model depicts the original structure as realistically as possible. For example, if cake is used in preparing an edible model of a microscope, the microscope should not be just an icing outline on top of the cake. Instead, the cake should be trimmed around the edges to depict the shape of the microscope. Edible materials such as candies should be added to represent such structures as the focusing knobs and mirror. Building up portions of the model with extra layers of icing imparts a raised or bas-relief effect.

In creating an edible model of the heart, red licorice sticks could be used for arteries and colored icings for the chambers. In a model of a cell, cherries could represent riobosomes and ribbon candy could represent the endoplasmic reticulum.

Worksheet 1.6–1 can be distributed to students at the beginning of the school year. Page 1 of the worksheet outlines the basic instructions for creating and presenting to the class an edible model. It also includes the criteria used in evaluating the edible model. Pages 2, 3, and 4 show pictures of actual edible models prepared by students. (Some copiers will reproduce pages 2, 3, and 4 effectively; others might not.)

Before the student presents an edible model to the class, you should take a 35-mm color slide of the model. At the end of the school year, slides can be shown of all edible models presented during the year. Using Worksheet 1.6–2, students can vote for the EMOY (Edible Model of the Year) award. You might want to give extra credit to the winner.

After you take a picture of the edible model, the student makes his or her presentation to the class. Following the presentation, the edible model can be shared with the class and consumed.

When serving the edible model, use clean paper plates or paper towels. Caution students not to let the food come into contact with lab, desk, or tabletop surfaces, as these surfaces are potentially unclean.

An article about edible models written, perhaps, by a student and submitted to a local newspaper (along with some photographs of the models) provides excellent public relations during the school year.

WORSHEET 1.6–1

Guidelines for Preparing
and Presenting an Edible Model

Preparing the Edible Model:

1. Your teacher will help you decide what picture to use and what structures to emphasize in preparing your edible model, as well as what information you are to present to the class.
2. Your model should be constructed out of edible materials only. Be creative in your selection. Cakes are not the only possibility. Consider such possibilities as pie, jello, pizza, cheese, and so on.
3. Your model should be prepared in the shape of the structure being depicted. (The structure should not just be an icing outline on the surface of a cake, pizza, etc.)
4. Candies and other edible materials can be used for individual component parts of a structure. For example, a cupcake or an apple could be used to represent the nucleus of a cell.
5. Extra layers of icing can be used to build up certain aspects of a structure (for example, the membranes and/or the walls of a cell).
6. No parts should be labeled on the edible model.

Presenting the Edible Model to the Class:

1. Your teacher will take a picture of your edible model.
2. Following the picture taking, walk around the classroom so that each student can get a close look at your model.
3. Next, present the information to the class. For example, point out each structure, giving its name and function or significance. (You must do this from memory; do not use your notes.)
4. Be certain that you can properly pronounce each name or term used in your presentation.
5. Upon completion of your presentation, cut and serve the edible model to the class.

Criteria Used in Evaluating Your Performance:

1. Prior to presentation, your teacher will examine your edible model for accuracy. Is it an accurate representation of the picture upon which it is based?
2. Were all names and terms pronounced accurately?
3. Was all information presented accurately to the class from memory?

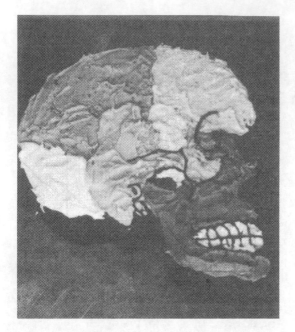

This edible model has been made using cake. Different-colored icings differentiate bones and markings of the skull. Note that the cake has been cut in the shape of the skull.

1. The Human Skull

This edible model has been made of cake and decorated with various candies representing the different sections of the fingers as well as the carpal bones of the wrist.

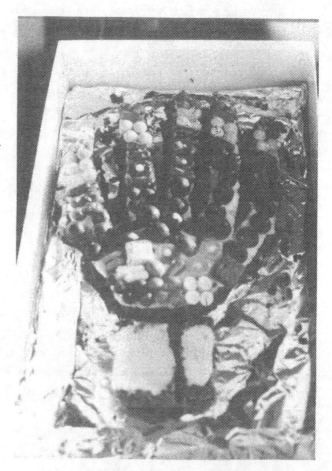

2. The Hand and Wrist

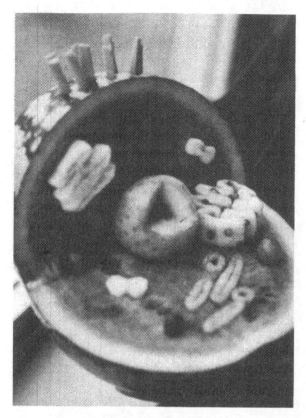

This edible model was made using fruits. The cell is watermelon. The nucleus is an apple, with icing dots on it representing pores through the nuclear membrane. Other structures are depicted using grapes, marshmallows, pineapple, and so on.

3. The Cell

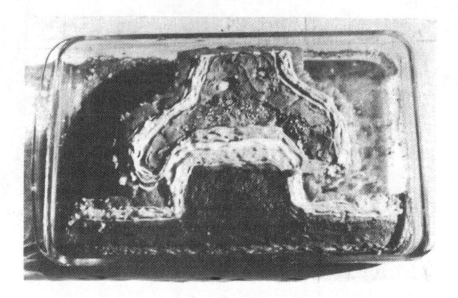

4. A Synapse

Using cake, colored icings, and chocolate jimmies, the basic structure of a synapse between the dendrite of one neuron and the axon of another is shown in detail.

5. The Human Eye

Using a brownie mix, colored icings, and various candies, the basic structure of the eye is illustrated. The outer sclera layer is icing, the middle choroid layer consists of miniature marshmallows, and the inner retinal layer is made of jelly beans.

6. Aerolar Connective Tissue

A long pasta noodle running from upper left to lower right depicts a collagenous fiber. Thin streaks of dark icing indicate elastic fibers. Small, round candies show some of the specialized cells found in the tissue.

Name: _____   Date: _____

Voting for the Edible Model of the Year

Instructions:

Your teacher is going to show the slides of all edible models presented during this school year. You will first view the slides without voting on them. This will give you an opportunity to do some relative comparing of the quality of the models.

During the second showing of the slides, you will be asked to rate each slide by giving it a number from 1 to 5, with 5 being the best. Use the rating form presented below on this worksheet. If there are more than 42 slides, add to this worksheet. The rating number that you give a slide is based on your opinion of how accurately that model depicts the structure upon which it is based.

All worksheets will be turned in, and the total score for each model will be determined. The model receiving the highest score is the winner and, therefore, the Edible Model of the Year.

RATING SLIDE #	RATING SLIDE #	RATING SLIDE #
_____ 1.	_____ 15.	_____ 29.
_____ 2.	_____ 16.	_____ 30.
_____ 3.	_____ 17.	_____ 31.
_____ 4.	_____ 18.	_____ 32.
_____ 5.	_____ 19.	_____ 33.
_____ 6.	_____ 20.	_____ 34.
_____ 7.	_____ 21.	_____ 35.
_____ 8.	_____ 22.	_____ 36.
_____ 9.	_____ 23.	_____ 37.
_____ 10.	_____ 24.	_____ 38.
_____ 11.	_____ 25.	_____ 39.
_____ 12.	_____ 26.	_____ 40.
_____ 13.	_____ 27.	_____ 41.
_____ 14.	_____ 28.	_____ 42.

1.7 VIDEO TAPE AND 35-mm SLIDE ACTIVITIES _____

Increasing numbers of students have access to videotape recorders, which provide excellent opportunities for them to share their creative endeavors through classroom projects. Although we discuss videotaped projects here, similar results can be achieved using 35-mm slides. Consider the following activities, which represent only a few of the many that can be approached using videotape:

Part-Time Work Opportunities: Many students have part-time jobs related to science and health. They may work in nursing homes and hospitals, or doctors' offices and veterinarian labs. Have the student obtain permission (you might wish to obtain it for the student) to tape a career-oriented job description of what he or she does. Other employees that the student works with can be interviewed on tape, and a tour through the work environment can be taped. This type of presentation to the class can be valuable in exposing students to the many different part-time job experiences awaiting them if they are so inclined. Of further value is that part-time job experiences (especially those in the fields of science and health) listed on a student's college application can add considerably to that student's chance for admission.

How-To Activities: There is almost no end to the number of how-to activities that can be videotaped. Examples include the proper way to use a microscope; the safe way to prepare agar; how to identify trees using the leaves; cloud type identification; the use of the laboratory balance; and how to prepare percent solutions.

Specific Topics or Concepts: Instead of preparing a written report (on diabetes, for example) to be read to the class, the student could prepare a videotape which could include descriptions of the disease supplemented with visual aids and interviews with afflicted individuals.

A student might select as a topic the operation of the local sewage treatment plant. Once he or she has obtained permission to visit the plant, a tape could be made explaining the various steps of water treatment. Interviews with the employees could further enhance the videotape presentation.

Career Profiles: The use of student-prepared videotapes exploring various career opportunities can be invaluable in providing guidance material for your entire class. A nursing career tape could include interviews with several different types of nurses, ranging from the operating room nurse to the hospital floor nurse. Educational requirements, job descriptions, and employment outlook opportunities could be included in the tape.

1.8 STUDENTS SETTING CAREER GOALS

It can be argued that the sooner a junior or senior high school student has a career goal in mind, the sooner his or her education takes on new meaning and relevance. It may be better to set any career goal whatsoever rather than have no goal at all. Goals can always be changed during high school or during college, but without any goal there is no direction for the student. A goal that has been established during high school allows the student to begin college with a clear-cut goal in mind. Moreover, with today's escalating college costs, it is more important than ever that each college course has significance for the student. Worksheet 1.8–1 provides a format for the student to use in articulating a career goal. If a student declares that he or she has no goal, insist that one be chosen. It would be helpful to have each student select and fill out a worksheet for at least two possible career choices, with at least one of them in a field of science.

An item that elicits considerable classroom interest is a poster display of the various careers that your students are currently interested in. Worksheet 1.8–2 is a form that you can use in gathering career information from your students. Once gathered, the information can be transferred to a large oaktag poster. The format of the poster can be similar to that of the worksheet. In addition, the poster can be made more eye-catching by letting a student add some colorful design to it.

The career poster should be updated at the end of each marking period. This is easily accomplished by handing out another copy of Worksheet 1.8–1, instructing that only those students who have made career changes fill it out.

You will be impressed by the number of students who scan the posters. Students are naturally interested in what career choices their peers have made. The poster displays help to keep students focused on career choices if they already have one, and encourage them to think about career possibilities if they do not have one. Further, the poster is an immediate source of information for you about the career choice of each student.

Consider, for example, a particular student who is having problems in a biology course and who has indicated a career choice of becoming a veterinarian. This is an ideal time to discuss with the student his or her career choice. The discussion could inspire the student to work harder toward a career in veterinarian science, or it could spur a change of career goals.

The career poster can serve as a guide for you in selecting classroom speakers. It can also be a guide for the speaker, who can become familiar with the students' career interests prior to talking to them.

1.9   THE BIOLOGY CLASSROOM CAREER RESOURCE CENTER __

The career resource center is a valuable adjunct to the classroom. Regardless of grade level, the student who has a career goal is likely to take learning more seriously. Education becomes relevant as a means of obtaining that goal.

The following are guidelines for setting up your own career resource center:

- Select an area of the classroom that has some wall space for bulletin board displays and room for chairs and a table (for displaying materials).
- Begin collecting pamphlets and brochures on all fields of health and health careers. Sources include doctors' and dentists' offices, hospitals, health agencies listed in the yellow pages, and display booths in malls and at career fairs. Your students can be encouraged to help collect materials.
- Borrow books and audiovisual materials from the school library to use in your classroom.
- Ask for student volunteers to gather materials and set up poster and tabletop displays featuring the career of the month.
- You might want to instruct the class to gather certain information from the resource center about each monthly featured career. You can use the information for testing purposes. For example, you can ask students to provide (a) the name of the featured career, (b) a brief job description, and (c) a job outlook. Testing students on career-of-the-month information assures that the work of student volunteers will be used by other students.

WORKSHEET 1.8–1

Selecting a Career Goal

1. Select what you consider to be your favorite possibility for a career goal:

2. List at least three specific reasons why this is your favorite choice:

    a.

    b.

    c.

3. Using resources of your guidance counselor and/or library, obtain the following information about your career goal:

    a. What particular personal skills are needed?

    b. What high school courses would be most suited to achieving your goal?

    c. What courses will you be taking during your first two years of college?

    d. What is the minimum number of college years of preparation?

    e. What is the future employment outlook?

    f. What is the starting salary?

    g. What are the opportunities for advancement?

    h. Will much of your job task involve working alone, or will much of it involve interacting with other workers?

    i. Will your job task include coming directly into contact with the public?

    j. Are there any aspects of this career that you might find undesirable?

k. Are any scholarship or foundation monies available for financial grants? (If so, list them.)

4. Interview a person who is working at the same career as your goal. The following questions are suggested as a guide for your interviewing. You will have many others, which should be listed on a separate sheet. Record all answers.

a. At what age did you first think that you might be interested in your chosen career?

b. What are some personal factors that made you think that you would like such a career?

c. What other people, if any, had an influence on your choice of a career?

d. Have you ever been sorry that you chose the career that you did?

e. What other careers, if any, do you think that you would have been happy in?

f. Could you list three positive aspects of your career choice?

(1)

(2)

(3)

g. Could you list three negative aspects of your career choice?

(1)

(2)

(3)

5. In the space below, summarize your feelings as a result of interviewing an actual person working in the field of your career goal.

Student Career Choices

Subject:

Class Section:

Instructions: Write your name in the left-hand column and the career that you think you are currently interested in in the right-hand column. If you do not have a career choice, please put a question mark in the career choice column.

STUDENT NAME   CAREER CHOICE

1.
2.
3.
4.
5.
6.
7.
8.
9.
10.
11.
12.
13.
14.
15.
16.
17.
18.
19.
20.
21.
22.
23.
24.
25.
26.
27.
28.
29.
30.
31.
32.
33.
34.
35.

1.10 THE BIOLOGY CLASSROOM READING CENTER: GETTING THE STUDENTS TO READ MORE _____

One way to encourage students to read more is to expose them to a wide variety of reading materials presented in an appealing and stimulating display. This can be accomplished by having a reading center in your classroom. You can obtain books for the reading center on short-term loan from the school library. (Library staffs are usually supportive in this endeavor, and may even offer to come into your classroom and present a program on the enjoyment of reading.) In addition, you can encourage students to bring in from home, on loan to the reading center, favorite books of their own. When provided with the opportunity, students are receptive to sharing personal favorite books with their peers.

The reading center need not take up a great deal of classroom space—it can consist of a table next to a wall with a bulletin board. If no bulletin board is available, oaktag sheets can be attached to the wall as a substitute.

The reading center can be organized around current topics being studied in class. For example, if the topic is reptiles, choose books on reptiles that you feel would be interesting to your students. As class study topics change, change the reading center book display.

Allow students to sign out books from the reading center. You will be pleased with the success of your endeavor. Many students do not have ample opportunity to get to the library during the course of the school day. The reading center successfully brings selected portions of the library to them.

The following are examples of reading center activities for students:

The Student Book Center Committee: This committee can be comprised of two or three students, who are assigned the responsibility of selecting reading center books for the current topic. They can also prepare a bulletin board type of display to enhance interest in reading about the current topic. Worksheet 1.10–1 will guide the committee in its activities. It is designed so you can remove the bottom portion and keep a record of books (and their copyright date) for your own personal use in (a) recommending books to students and (b) ordering new and updated books when necessary.

The students on the committee can be changed for each topic. All students involved with the book center can be awarded extra credit.

The Student Book Review: Interested students might wish to prepare and present book reviews to the rest of the class. Worksheet 1.10–2 instructs students in preparing and presenting book reviews.

WORKSHEET 1.10–1

The Student Book Center Committee

Instructions for setting up the book center display:

1. Determine the topic for display (ask your teacher).
2. Select eight to ten books from the library that deal with the topic.
3. Arrange the books in an attractive display on the reading center table.
4. Prepare an eye-catching bulletin board type of display dealing with the topic. Such a display might include the following:
   - Title (printed, or letter cutouts) announcing the display topic
   - Pictures from magazines (Have your teacher or librarian make copies of the pictures. Do not cut pictures out of magazines unless told it is permissible.)
   - Colored sketches and drawings of your own
   - A list of occupations and careers associated with the topic
   - Names of committee members responsible for the display.
5. Fill out the form at the bottom of this sheet prior to turning in the worksheet to your teacher.

Topic:

List of Textbooks Selected for the Display:

	NAME OF TEXTBOOK	COPYRIGHT DATE
1.		
2.		
3.		
4.		
5.		
6.		
7.		
8.		
9.		
10.		

Name: _____   Date: _____

The Student Book Review

Instructions: Use the following information as a guideline in preparing and presenting your book review.

1. Select a book that you are truly interested in reading and reviewing.

2. As you are reading the book, jot down notes from time to time dealing with the following:

   • The overall scope of the book (*Example:* The book dealt solely with poisonous snakes of North America.)

   • Particular parts of the book that especially interested you, and why (*Example:* Snakes of the desert, because of the unique ways they have of adapting to desert conditions)

   • Specific pieces of information to relate to the class (*Example:* The rattlesnake referred to as the sidewinder gets its name due to . . . .)

3. As you prepare your book review for presentation to the class, the most important factor to keep in mind is obtaining and maintaining your audience's attention. You will bore your audience if you just stand in front of the classroom and read a written book review. Your goal is to make your subject interesting to your classmates, and to stimulate them into wanting to read books on topics that interest them. To achieve this, consider using the following visuals:

   • Pictures shown on the classroom screen using an opaque projector (Refrain from passing pictures around from student to student. It is much more effective to draw their attention to one picture at a time on the screen. This allows you to comment on the picture and take questions.) (*Example:* Show a picture of a rattlesnake and discuss the question: How do rattlesnakes avoid the very hot desert temperatures?)

   • A colorful oaktag poster that introduces your topic (*Example:* The poster might contain the name of the book, with highlights from the table of contents. Further, it could include sketches and/or pictures of snakes.)

- A map showing the major deserts of the world (*Example:* The map can show desert areas of the world about which you plan to read in the future. Thus what began as a general interest in reading about poisonous snakes of North America has developed into an interest about desert snakes of the world.)

4. You do not have to memorize your presentation word for word. You can always refer to your notes for guidance.

5. Make sure that you can pronounce all words and terms that you are going to use in your presentation. If you need assistance, check with your teacher.

WORKSHEET 1.10–3

The Student Book Examination

Instructions: This activity involves reading a book, taking notes on your reading, and taking an oral or written test. Use the following information as a guideline in accomplishing this activity.

1. Select a book that appeals to your interests, and make sure that your teacher approves of it.
2. As you read the book, record notes (along with the page numbers on which the information is found) for all information you wish included in your test. The space at the bottom of this worksheet is designed for this purpose. At least twenty pieces of information should be recorded. (If you need more space, use the back of this worksheet.)
3. Your teacher may ask two or three additional questions. These questions will be general in nature and will show your teacher that you have read the entire book.

INFORMATION + PAGES	INFORMATION + PAGES
1.	11.
2.	12.
3.	13.
4.	14.
5.	15.
6.	16.
7.	17.
8.	18.
9.	19.
10.	20.

The Student Book Examination: This activity is for any student who would like to read a book and take notes on interesting things he or she is learning from the book. The student is given an oral or written test on his or her notes, and the grade received is used as extra credit. You may want to add two or three questions of a general nature to help ensure that the student has read the entire book.

Worksheet 1.10–3 guides students through this book examination activity.

1.11   THE STEPS TO AN A GAME

The game board can be easily constructed from a few basic materials. You could offer extra credit to a student who volunteers to construct the game board. Worksheet 1.11–1 shows what the fully constructed game board will look like. (*Note:* The basic construction plan can be modified according to your own desires. For example, you can alter the dowel size, game board size, and so on.)

A.   Assembling the game board frame:

- Refer to Worksheets 1.11–2 and 1.11–3 as you proceed.
- Obtain the pieces of wood for the bottom and the top of the game board frame (Worksheet 1.11–2).
- Drill five holes (make certain the dowels illustrated on Worksheet 1.11–3 will fit through the holes) in the top piece of the game board frame as follows:

The first hole is drilled 6" from the left-hand end of the piece of wood.
The second hole is drilled 7" to the right of the first hole.
The third, fourth, and fifth holes are each drilled at 7" intervals to the right.

- Along the bottom piece of the game board frame, make five countersinks corresponding in position to the five holes drilled in the top piece.
- Obtain the pieces of wood for the sides of the game board (Worksheet 1.11–3).
- The top, bottom, and sides of the frame can now be assembled using nails (along with wood glue, perhaps).
- Cross supports at the corners might be needed for increased frame stability.

B.   Assembling the game board boxes:

- Refer to Worksheet 1.11–3 as you proceed.
- Obtain the pieces of wood needed for the five game board boxes.
- Drill a dowel hole in the top and bottom pieces for each box.
- Assemble each of the boxes using nails and/or wood glue.

The Fully Assembled Game Board

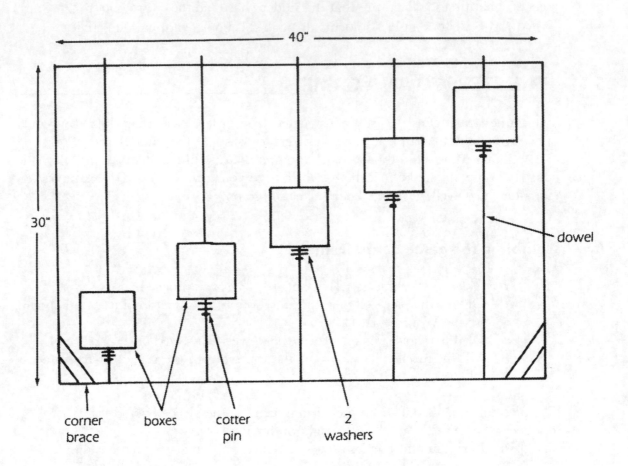

The Top and Bottom of the Game Board

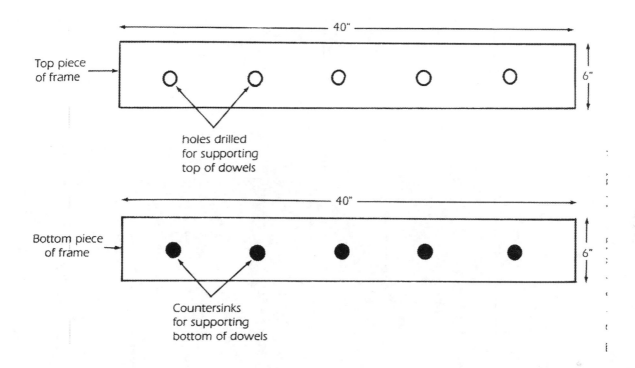

The Sides, Boxes, and Dowels for the Game Board

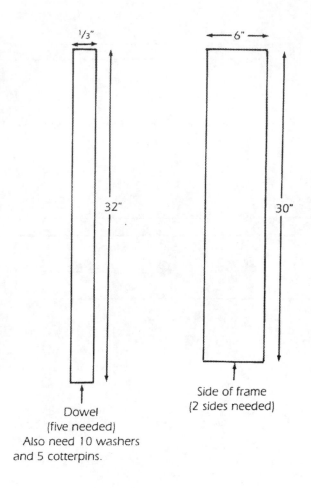

1/3"

32"

Dowel
(five needed)
Also need 10 washers
and 5 cotterpins.

6"

30"

Side of frame
(2 sides needed)

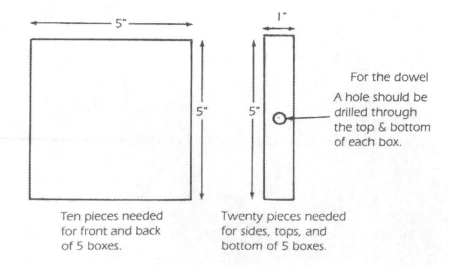

5"

5"

Ten pieces needed
for front and back
of 5 boxes.

1"

5"

For the dowel

A hole should be
drilled through
the top & bottom
of each box.

Twenty pieces needed
for sides, tops, and
bottom of 5 boxes.

C.   **Preparing the dowels:**

- Obtain five dowels (illustrated on Worksheet 1.11–3).
- Drill a hole through each dowel at the level at which it is going to support a box. (Note on Worksheet 1.11–1 that each dowel supports a box at a different level.)
- Secure a cotter pin through each drilled hole.
- Slip two washers over each dowel and allow them to rest on the cotter pin. (This will allow the box to be rotated easily.)

D.   **Assembling the boxes on the dowels:**

- Pass a dowel through the bottom and top openings of each box, allowing the box to rest on top of the washers.

E.   **Assembling the dowels (with boxes attached) to the game board frame:**

- Pass the upper portion of each dowel up through the hole in the top of the game board frame.
- Secure the base of each dowel in the corresponding countersink.
- The game board should now be entirely assembled.

The following are ways to enhance the completed game board:

- Add a plaque naming the student(s) responsible for construction.
- Paint the game board.
- Add a string of holiday-type lights to the game board.
- Have a student assemble a doorbell-type buzzer to denote incorrect answers.

The rules (which you can modify) for playing "The Steps to an A Game" are as follows:

1. Students play as teams of two. (If you have several teams wishing to play, they can wait their turn quietly in the hallway.)
2. Write a different task on each of five sheets of paper. Examples of several different tasks are: Define Zoology; Name bones that comprise the human skull; Name the layers of the eyeball; Define aseptic technique. The general theme of the tasks will depend on classroom material currently being studied or reviewed.
3. Tape the sheets of paper, with the tasks written on them, out of view, on the back of the game board boxes.

4. Position the game board on a desk at the front of the classroom.

5. Position two chairs, one with its back to the game board, facing each other in front of the game board.

6. Seat the two student contestants in the chairs.

7. The student who faces the game board is referred to as the giver, and the student who does not face the game board is referred to as the receiver.

8. During game play, the giver sees the task asked for, such as: Name bones that comprise the human skull. The giver then begins to give the names of some skull bones to the receiver.

9. During the game, the receiver listens to the information that the giver provides and attempts to identify the task, which in this case is the names of bones of the human skull.

10. Game play progresses from the lowest game board box to the highest.

11. Game play begins when a student helper, positioned behind the game board, turns the lowest box around, revealing the task to the player who is the giver. If the box is correctly solved by both the giver and the receiver, it is left as it is. If the box is not correctly solved by the giver and the receiver, it is turned back around.

12. If either the giver or receiver experiences difficulty at a box, that box may be passed and returned to later.

13. The game has a time limit, for example 2½ minutes.

14. If the game is won within the time limit, both students receive an A. If not, then neither student receives extra credit.

15. The game is won if the receiver correctly identifies the tasks on all five game board boxes, and the giver has provided accurate clues for all five game board boxes.

16. The teacher will be the judge of the accuracy of the giver's and receiver's responses.

17. During the course of the game, if at any box the giver uses a word included in the task description, the game is ended.

It is likely that "The Steps to an A Game" will become a favorite of students. It is a fun way to test student knowledge, and it can be used to motivate students to study at times other than for tests.

If you decide to use the extra-credit grade of an A as a test grade equivalent, this will provide strong motivation for student participation.

1.12 THE STUDY CLINIC MINICOURSE

Whenever a student is not achieving academically in a course, he or she needs to know that help is available.

If a student is having difficulty learning the materials presented in a course, it is usually due to one of the following factors:

a. The student is not using proper study techniques.
b. The student is not interested in the material.
c. The student is experiencing problems in his or her personal and/or home life.

All three of these factors demand that you give the student personalized attention. The study clinic minicourse provides an opportunity to do so. Although the minicourse is designed to help a student who is not using proper study techniques, if there is a lack of student interest or personal problems are involved, you can address them in the context of the minicourse.

The personalized attention given during the minicourse is often the catalyst for renewed student effort. In addition, through individual and small group sessions, you and the student(s) can get to know each other on a less formal basis.

Your free time is always at a premium. Ideally, you would be available after school each afternoon for student help. Obviously, this is not always the case, with department meetings, grade level meetings, faculty meetings, and so on. Therefore, you might wish to set aside one or two afternoons a week solely for the purpose of study clinics, perhaps on Tuesdays and Thursdays. Students would have a choice of which day to attend.

Students can use Worksheet 1.12–1 in diagnosing deficiencies in note taking and/or studying. This worksheet is the initial task students should carry out when attending the study clinic for the first time. You should work along with students as they complete the worksheet.

The rest of the format for the study clinic should be based on the study techniques discussed in Worksheet 2.4–3.

1.13   DEVELOPING YOUR OWN COURSE HANDBOOK FOR STUDENTS

Preparing a course handbook for your students can be of great value in clearly articulating topics ranging from the frequency of quizzes and tests to the specifics of doing extra-credit work. Worksheet 1.13–1 presents an example of a student handbook table of contents for a biology course in human anatomy and physiology.

The course handbook is a valuable tool for students taking the course, students who might be interested in taking the course, and parents and guardians. The course handbook imparts a note of seriousness to the course; it shows that the course is a result of forethought and planning, with the student's welfare in mind.

The following is an example of sections of a handbook for a course in anatomy and physiology. It can be modified for any course you are teaching.

Introduction to the Course:

"This is a course in human anatomy (structure) and physiology (function), designed for those students who might wish to plan for a career in one of the medical or allied health fields as well as those students who just wish to learn more about the human body. The course includes a considerable amount of studying, involving memorization of body structures, and it involves a suggested minimum of 30 minutes an evening of homework, including completing textbook correlated worksheets and learning materials covered in minilectures and classroom lab activities."

Content Outline of the Course:

Include here an overview of topics studied during the course.

WORKSHEET 1.12–1

Initial Problem Diagnosis and Remedy

This worksheet is designed to help you pinpoint the source of your study problem. Your teacher will be working directly with you as you complete it.

You need the following materials:

- Textbook
- Your most recent test paper
- All notes that you used in studying for the above test
- Note paper

Question 1

Are the correct answers to the questions that you missed on the test in your notes?

Procedure for Answering Question 1:

a. On a separate piece of paper, write each question from the test that you either answered incorrectly or left blank.
b. Check your notes for the correct answer to each question. If you have the correct answer in your notes, write a YES by that question; if you do not, write a NO.

Diagnosis:

- If you found questions for which you had incorrect answers in your notes, you are not recording the correct information in your notes.
- If you found questions for which you had no answers in your notes, you are not getting the information in the first place.
- If you had all of the correct answers in your notes, you are not thoroughly studying and memorizing the materials.

Remedy:

- If you are not recording the correct information in your notes, you should listen more closely in class, and work much more slowly when filling out any worksheets based on text material.
- If you are not getting the information in the first place, you are not taking the required notes in class and/or doing assigned worksheets.
- If you are not thoroughly studying and memorizing the materials, then you need a technique that will help you.
- Attending the next few study clinic sessions will be of great help to you in preparing for future tests.

Student Handbook Table of Contents for a Course in Human Anatomy and Physiology

The following is an actual table of contents from a course in anatomy and physiology, and provides a format that can be modified easily for any biology course.

Dissections Carried Out in the Classroom:

"The following dissections are performed by students in groups of two (you can work individually if you desire):

a. The rat: an overview of major systems, body cavities, and general overall body structure plan
b. The cat: detailed study of the major systems; selected joints and muscles.
c. The calf or sheep brain
d. The calf or sheep eye
e. The calf or sheep heart.

IMPORTANT NOTE: If you anticipate any personal problems with regard to dissection, please discuss it with the teacher."

The Study Clinic Minicourse:

(This is discussed in Section 1.12.)

"When a student is having difficulties in a course, the reason is usually not a lack of ability on the part of the student, but rather less than desirable study habits and techniques. Therefore, Tuesdays and Thursdays after school have been set aside for a study techniques clinic. You might wish to attend one or both, preferably on a weekly basis, until your grade improves. This clinic is literally your key to obtaining the highest grade that you choose in this course."

Frequency of Quizzes and Tests:

"There are no quizzes in this course. There are a minimum of six exam grades (three from text material and three from lab practicals) per nine-week marking period. All tests will be announced at least four days in advance."

Making Up Missed Tests:

"All tests missed during a marking period will be made up at the end of that marking period on a date to be announced as the marking period comes to a close."

The Grade Option:

"If, at the end of the first marking period, your grade average is a B or better, you can take advantage of the grade option. This means that if you have library work to do, or work to do with another teacher, a pass will be written for you excusing you from class. The grade option can be used as long as your yearly average remains at a B or better. Obviously, there will be class activities such as dissections, films, and speakers during which the grade option will not be in effect."

The Voluntary Oral Exam:

This is discussed in Section 1.4.

The Steps to an A Game:

This is discussed in Section 1.11.

The Pic-a-Tag Technique:

This is discussed in Section 1.2.

Extra Credit: The Disease of the Week:

Worksheet 1.13–2 is the handbook format describing this extra-credit activity. The topic a student selects must correlate with the system of the body being studied at that time. For example, the disease epilepsy correlates with a study of the nervous system.

Extra Credit: The Occupation Presentation:

Worksheet 1.13–3 is the handbook format describing this extra-credit activity. Whenever possible, the occupation should correlate with the system of the body being studied. For example, orthopedics correlates with the skeletal system. The occupation presentation activity exposes students to a wide variety of potential future careers.

Extra Credit: Teaching the Class for a Period:

"If you wish to teach the class a lesson on a particular topic that interests you, have the topic approved by your teacher. Your teacher will then give you a proposal form to complete that itemizes your presentation in detail."

Worksheet 1.13–4 is the proposal form that students receive for the preceding activity.

Extra Credit: The Edible Model:

This is discussed in Section 1.6.

Extra Credit: Further Ideas:

In addition to the extra-credit activities already discussed in this handbook, the following can be considered:

Name: _____   **Date:** _____

Extra Credit: The Disease of the Week

Procedure for preparing the report:

1. The disease must relate to the current material being studied in class.

2. The format of your paper should include the following headings:

    a. cause of the disease

    b. parts of the body affected

    c. how the parts of the body are affected

    d. how the disease is treated

    e. effectiveness of the treatment

    f. how widespread the disease is

    g. age group or type of person most likely to get the disease.

3. At least one visual aid should accompany your report. It should be clear and easily understood.

4. The minimum length of the report should be the equivalent of three full pages of handwritten, single-spaced material.

5. A bibliography page should be included.

6. Give the rough draft copy of your report to the teacher at least five days prior to giving the report in class. Your teacher will return your reviewed rough draft to you along with ditto masters on which to prepare your final draft.

7. Return your final draft to your teacher at least two days prior to giving your report in class. This will give your teacher time to prepare copies of your report for handing out to the students.

8. You will present your report orally to the class. Using an opaque projector, you will also show and explain all visual aids.

9. Be certain that you can correctly pronounce and define all terms used in your report.

WORKSHEET 1.13–3

Extra Credit: The Occupation Presentation

Procedure for preparing the occupation presentation:

1. If desired, two students can work together provided that they share equally in the preparation and class presentation.

2. Fill out the form at the bottom of this worksheet and turn it in to your teacher.

3. The occupation presentation includes

    a. a written report

    b. an oral class presentation

    c. a bulletin board type of display.

4. The report should include the following headings:

    a. job description

    b. available colleges

    c. education needed (how many years of college; type of courses taken; internship, and so on)

    d. available financial college aid

    e. employment outlook

    f. salary.

5. At least five days prior to class presentation, turn in your rough draft copy of the report. After reviewing it, your teacher will return it to you along with ditto masters.

6. After preparing your final draft on ditto masters, turn it in to your teacher at least two days prior to class presentation. Your teacher will prepare copies of your report to hand out to the students.

7. The class presentation consists of going over your report and the bulletin board type of display.

_____   _____   _____   _____

Occupation Presentation

Student Name(s):

Occupation Topic:

Name: _____   **Date:** _____

Proposal for Teaching the Class for a Period

Complete the following information and submit it to your teacher:

1. Name of topic: _____

2. An outline of the topic, identifying the main areas of information that you plan on presenting to the class (Use additional paper if necessary.)

3. List any visual aids that you plan on using:

   a.   e.

   b.   f.

   c.   g.

   d.   h.

4. List any pieces of audiovisual equipment that you will need:

   a.   e.

   b.   f.

   c.   g.

   d.   h.

5. Prepare on a separate piece of paper a quiz of at least 10 questions that you can use to evaluate student learning. (The questions should represent what you consider to be the 10 most important pieces of information that you presented.)

> Papier-mâché models of organisms and structures studied.
> Preparing a science program to present to elementary students.
> Constructing and designing a puppet program for elementary students.

Your teacher will consider any unique idea that you come up with for extra credit.

The Student Progress Form for Parents:

Part A of Worksheet 1.13–5 is the handbook example of a form that can be sent home to parents and guardians in the event of, for example, an unusual grade average drop for a particular student. Or, you might want to use the form whenever a student's grade drops to the "danger zone" of a D.

The Chronic-Missing-of-Tests Form for Parents:

Part B of Worksheet 1.13–5 is the handbook example of a form that can be sent home to parents and guardians when the student has a habit of being absent on test days.

Detentions: Factors Involved in Assigning:

The following factors are usually involved when a detention is assigned to a student:

- Being late to class without a pass
- Not having the required materials for the day, such as notebook, textbook, colored pencils, and so on
- Doing homework for another class
- Attempting to catch some "shut eye," such as during a film or tape
- Any disruption, such as talking while classroom instruction is in progress.

The Film Program:

If you have a selection of films and tapes that you use on an annual basis, they can be listed in the handbook.

The Speaker Program:

If you have speakers that come in on a yearly basis to talk with your students, they can be listed in the handbook.

A good speaker program is an excellent adjunct to the classroom program. Speakers in all fields of biology can be obtained easily using the following sources:

1. Local hospitals
2. Local community colleges and universities
3. The Yellow Pages, under specific headings such as "The Leukemia Society of America."
4. Parents, through your students.

Teacher-to-Parent Forms

Part A: The Student Progress Form for Parents:

Dear _____:

This note is to inform you that _____ grade average has dropped below a C. His/her current grade average is: _____.

I have recommended that my Study Technique Clinic be attended until the grade improves, and would appreciate your reinforcement in this regard.

I have detailed below all grades received thus far in the school year:

GRADES:	Marking Period 1	Marking Period 2	Marking Period 3	Marking Period 4
1				
2				
3				
4				
5				
etc.				

If I can be of further help in answering any questions, please do not hesitate to call me at _____ (phone number).

Sincerely,

Part B: The Chronic-Missing-of-Tests Form for Parents:

Dear _____:

This note is to let you know that _____ has missed _____ tests thus far this marking period. I bring this to your attention because students often are not as well prepared for the make-up test as they would have been for the original in-class test. As a result, grade averages could suffer.

If I can be of any further help, please call me at _____.

Sincerely,

The Anatomy Award Program:

This is discussed in Section 1.20.

A Parting Message:

The handbook can be personalized by adding a message that is perhaps inspirational in nature. Worksheet 1.13–6 is an example of such a message.

1.14   ACTIVITIES IN CREATIVITY

It can be fun to run a creativity contest for your students. You will be impressed with the creative ingenuity students display as they let their imaginations run wild with innovative thinking.

Before conducting any creativity contests, you might want to engage students in a discussion of creativity by posing some questions. For example, how does one define the word *creative?* How does one know if one has been creative? Are there any unique characteristics of the so-called creative individual? Are there any ways in which one can become more creative? (*Note:* Psychology texts are excellent resource materials for information on creativity.)

The following are descriptions of three creativity contests. As an incentive, you might wish to award extra credit to contest winners while awarding less extra credit to all contest participants.

Contest 1—Soap Carving: Two large bars of soap (Ivory® brand works well due to its soft consistency) are used by each team of two students. The teams are instructed to have one member carve a bar of soap into an object that will complement the object carved by the other member. For example, one member might carve a television set while the other member carves a VCR to go with it. One member might carve a shoe while the other member carves a foot to fit the shoe. A carved car might go with a carved trailer. The more imaginative and unique the creation, the better.

At the conclusion of the activity, another class judges the creations.

Worksheet 1.14–1 outlines the specific instructions for the contest, so distribute copies to student participants a few days in advance of the activity. This will allow the students time to develop some creative ideas.

SAFETY NOTE

For carving purposes, a laboratory spatula with a flat end is recommended. When used with caution, it will not injure the fingers or hand as something sharper, like a razor blade, is likely to do.

Worksheet 1.14–2 outlines the specific instructions for judging, so distribute it to the judges selected from another class. (*Note:* This worksheet form can also be used in judging contests 2 and 3.)

After judging has been completed, add up the total rating scores for each entry. Barring a runoff in the event of a tie, the team with the highest score is the winner.

An Example of an Inspirational Message That Could Be Part of the Course Handbook for Students

THE PROBLEM

It can be elusive, without a doubt,
a cerebral crisis, no way out.
    A real challenge for the positive of mind,
a Waterloo for the defeatist kind.
    A formidable scar suspended overhead,
its precipitous presence a bearer of dread.
    Yes, this it can be and much more too,
yet who is the master, it or you?
    The answer to this is in your belief,
your innate being, for supremacy or grief.
    You believe you are master then you are sublime,
you believe you are defeated, it's the end of the line.

M. Fleming

Name: _____   Date: _____

A Soap Carving Creativity Contest

Instructions:

1. Choose a student to work with you on this activity.
2. You will be given two bars of soap, one for each of you to carve.
3. Carving will be accomplished by using the laboratory spatulas provided by your teacher.

SAFETY NOTE _____

Even though the carving end of the spatula is not sharp, it can cause injury if not used with caution.

_____

4. You will have one class period in which to complete your contest entry.
5. The objects carved by you and your team member must complement each other in some way. (For example, if you carved a needle and your team member carved a spool of thread, these two objects would complement each other.)
6. If a piece of soap breaks during or after carving, toothpicks can be used to secure the broken pieces.
7. Be as creative as you can in this contest by carving objects that you think no one else would have thought of or imagined.
8. Upon completion of this activity, obtain a manila folder on which to place your objects for display.
9. Fill in the following form. Since student names might influence judging, fill in the code blank with a code of your own choosing. Remove the form along the dotted lines and tape to the surface of the manila folder for displaying and judging purposes.
10. Carefully transport your creation to the area of the classroom designated by your teacher.

— —

Code: _____.
Class name: _____.
Class section: _____.

THE NAME WHICH YOU THINK BEST DESCRIBES YOUR CREATION:

_____

Name: _____   Date: _____

Form for Judging the Entries in the Creativity Contest

Instructions:

As you judge each entry for uniqueness and creativity, you will be using a 1 to 6 scale; 1 is the poorest rating an entry can get and 6 is the best. Each entry is judged independently of the others. This means that more than one entry could receive the same rating (for example, a 6).

As you judge the entries one by one, fill in the following form:

TITLE OF ENTRY	RATING	TITLE OF ENTRY	RATING
1.		11.	
2.		12.	
3.		13.	
4.		14.	
5.		15.	
6.		16.	
7.		17.	
8.		18.	
9.		19.	
10.		20.	

Additional information:

Explain in as much detail as possible what criteria you used in judging the entries. While it might be difficult to say why you rated an entry a 1 instead of a 2, or a 3 instead of a 4, it should be considerably easier to say why an entry was rated as a 6 rather than a 1.

After announcing the winner, discuss with your students the criteria listed by the judges (at the bottom of Worksheet 1.14–2) that influenced their rating of the entries. The main point the students should understand is that the judging of any contest of this nature is highly subjective. In other words, when it comes to judging creativity, it is extremely difficult for people to agree on just what makes something creative.

Contest 2––Toothpick Constructions: One box of flat toothpicks and a tube of fast-drying cement are used by each team of two students. The teams are instructed to create what they consider to be a creative toothpick construction. Examples of such constructions are a Ferris wheel that turns, pieces of furniture, and impressionistic sculptures.

Worksheet 1.14–3 outlines the specific instructions for the contest, so distribute it to the students a few days in advance of the contest.

Variations of this contest might involve the following:

- Using plastic toothpicks or small, plastic-coated straws along with quick-drying plastic cement
- Having students assemble their constructions at home with no restrictions on the numbers of boxes of toothpicks used (Some very impressive constructions will result.)
- Having students design and assemble similar constructions, such as a bridge, with the winner being the bridge that supports the greatest weight.

Contest 3––Creative Drawing: Each participating student receives one copy each of Worksheets 1.14–4, 1.14–5, and 1.14–6. You will notice that each worksheet (contest entry) has a different figure drawn on it. These three contest entries are to be completed by the end of the class period.

Instruct each student to use a pencil and create one drawing on each worksheet using the figure already on the worksheet as a part of the drawing. Tell students to be as creative as possible in incorporating the worksheet figure into their drawings. On each worksheet, instruct students to write a personal code identification instead of their actual name.

Have students from another class judge the drawings using Worksheet 1.14–2 as a guide.

Variations on this contest might involve the following:

- Using colored pencils and/or crayons
- Being able to cut and/or fold the creative drawing as well as being able to glue or tape on added pieces of paper to give a three-dimensional quality
- Designing a get-well card (or other type of greeting card) complete with drawing and verse.

A Toothpick Construction Creativity Contest

Instructions:

1. Choose a student to work with you on this activity.

2. You will be given a box of toothpicks and a tube of cement to work with. These are the only materials allowed.

3. Be as creative as you can in your toothpick construction design.

4. You will have two class periods in which to complete your construction.

5. It might be advantageous to assemble individually the main pieces of your construction during the first class period. Then allow the cement used on these individual pieces to set overnight before cementing them together to complete your construction.

6. Following the completion of your construction, obtain a manila folder upon which to display it.

7. Fill in the form at the bottom of this sheet. Since student names might influence judging, fill in the code blank with a code of your own choosing. Remove the form along the dotted line and tape it to the surface of the manila folder for display and judging purposes.

8. Carefully transport your construction to the area of the classroom designated by your teacher for display.

Code: _____.
Class: _____.
Section: _____.

THE NAME WHICH YOU THINK BEST DESCRIBES YOUR CREATION:

_____

Code: _____.   Date: _____

Class: _____.

Section: _____.

WORKSHEET 1.14–4

Creative Drawing Contest Entry 1

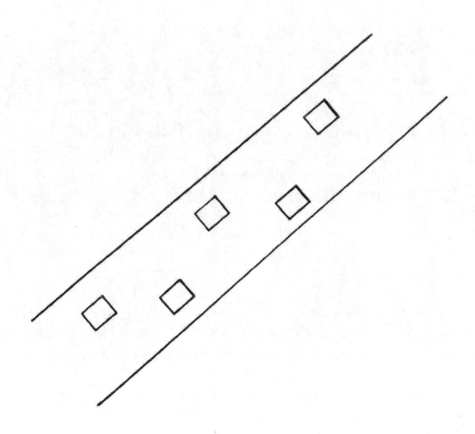

Code: _____.   Date: _____

Class: _____.

Section: _____.

WORKSHEET 1.14–5

Creative Drawing Contest Entry 2

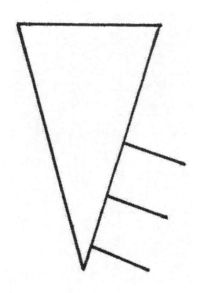

Code: _____.   Date: _____

Class: _____.

Section: _____.

WORKSHEET 1.14–6

Creative Drawing Contest Entry 3

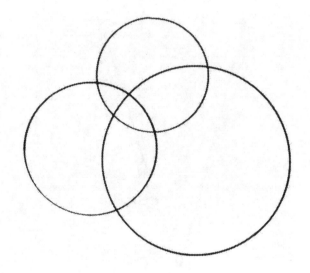

1.15 THE CLASSROOM INVENT-A-FAIR: A SCIENCE FAIR ALTERNATIVE

For those teachers who would like an alternative to the tremendous amount of work that goes into a full-fledged science fair, the Classroom Invent-a-Fair could be just the thing. It is a fun event, can be held in the classroom, and can be carried out with a minimum of teacher effort and preparation. It requires real ingenuity and creativity on the part of the participants, in many ways more so than the typical science fair project.

The Classroom Invent-a-Fair can be limited to students in just one class or can involve students from several classes. It can even involve different classes (each working as a unit) competing against other classes, each working as a unit.

The following steps outline the basic format for a Classroom Invent-a-Fair:

1. Distribute Worksheet 1.15–1 to the students. It provides an introduction to the Classroom Invent-a-Fair and a form that the volunteers who wish to participate must fill out.
2. Set a date for all prototypes to be turned in and for the actual event to take place.
3. Be willing to help students who might need assistance with such things as locating certain materials.
4. You may want to arrange for after-school get-togethers during which the "young inventors" can troubleshoot problems with each other as they work on their prototypes.
5. Select judges for the activity. These judges might be a combination of science teachers and students as well as qualified people from outside of the school.
6. Decide on prizes or awards. The science department budget may have some monies available to cover costs.
7. Consider having a local newspaper cover the event. This is excellent public relations.

Worksheet 1.15–2 is an example of a format that students can fill in and turn in with the prototype. These worksheets should be placed beside the prototypes and read by the judges as they judge the entries.

Because there are so many variables involved in judging projects such as these, a meeting should be held with the judges for the purpose of designing an evaluation format. Worksheet 1.15–3 is an example of a rating format.

1.16 ORGANIZING DAILY LESSON MATERIALS: THE MANILA FOLDER APPROACH

Organizing lesson materials, modifying lesson materials, adding new lessons for classroom trial and deleting old lessons that proved to be less than effective, storing lessons, and retrieving lessons on a yearly basis—all of these activities can be taxing and time consuming. This is especially true if you do not have some system of lesson material management. The manila folder approach allows you to keep your daily lesson materials within easy access so you can accomplish all of the aforementioned activities with a minimum expenditure of valuable time.

WORSHEET 1.15–1

The Classroom Invent-a-Fair:
An Introduction and Participation Form

Student Introduction

How many times have you said to yourself, "What an idea. I wish I had thought of that." Well, now is your opportunity to "think of that."

The classroom invent-a-fair is an activity that calls upon your ingenuity and creativity. You invent something entirely new, or add an innovative touch to something already existing. An example of the former might be designing and constructing a prototype of a device that would allow one to manually clean snow and ice, which might build up while driving, off the windshield of a car without the driver having to get out of the car. An example of the latter might be designing and constructing a model of a typewriter or word processor keyboard that would be easier to use than the standard one.

Whatever you design, it should be original and to the best of your knowledge not already in existence. Your design should be followed by the construction of a working prototype or model.

Remember, items such as the paperclip, clothespin, pencil, mousetrap, and traffic light all began as ideas in people's minds. Your idea may require extensive thought on your part, or it may be an idea that you already had at one time or another. It is the constant search for "a better mousetrap" that brings to fruition new and novel inventions. It's listening to people when they say, "I wish someone would invent a device that . . . ." It is important that you realize that you are just as capable as the next person of coming up with a new invention.

As a volunteer in the classroom invent-a-fair, you must follow four basic steps:

1. Originate an idea.
2. Discuss your idea with your teacher.
3. Build a working prototype.
4. Present it at the classroom invent-a-fair.

If you wish to challenge yourself to some really innovative thinking, fill out the form below and give it to your teacher.

CLASSROOM INVENT-A-FAIR ENTRY FORM

Name:

Idea:

What do you think is the value or importance of your idea?

Note any help that you think you might need from the teacher:

Name: _____ Date: _____

Official Student Entry Form to Be Used by the Judges

Student Instructions: Fill in this entry form with as much detail as possible. The judges are going to read it carefully as they evaluate the prototype of your invention. (Use extra paper if necessary.)

1. Name of your invention:

2. Discuss why you think your invention is important, noting its value and potential uses:

3. Discuss any difficulties or obstacles that you had to deal with and overcome in designing and building your prototype:

4. Discuss the basic construction of your prototype as well as how to use it (the judges will be operating the prototype):

A Sample Format for Judging Student Prototypes

Instructions: The judges, working together, should prepare this final summary, which will be given to the student. Ratings are accomplished using a 1 to 10 scale, with 10 being the highest rating.

1. Entry number:

2. Name of prototype:

3. Comments on the general merits of the prototype (its value and uses):

Rating: _____

4. Comments on the difficulties involved in designing and constructing the prototype:

Rating: _____

5. Comments on the actual design and usability of the prototype:

Rating: _____

Total Rating: _____

To implement the manila folder approach, you must do the following:

- You need one manila folder per each day's lesson materials.
- On the cover of each manila folder, you will need to record certain relevant information, such as that shown in Worksheet 1.16–1. Information recorded under the heading "ANY IMPORTANT ADDITIONAL NOTES?" is especially relevant in terms of keeping your lessons up to date. For example, you might note that a particular lab activity needs to be rewritten for added clarity, or perhaps deleted or replaced with another activity.
- If your school system requires that you keep a list of learning objectives for each lesson, keep them in the manila folder.
- Store the manila folders in numbered order on shelves under individual chapter or unit headings.

There are two additional major benefits of the manila folder approach:

- There will be no lull in classroom activity. If you are finished with one folder and have 10 minutes to spare, move on to the next folder.
- Administrators and department heads will be impressed when you tell them that you have a manila folder complete with learning objectives for every day of the entire school year.

1.17   THE STUDENT CUMULATIVE FOLDER _____

Keeping a cumulative folder throughout the school year for each student can be valuable to you and to students and their parents or guardians.

When you return to the class the first quiz or test papers of the year, give each student a manila folder. Have students put their name and class name and section on the front cover of the folder. Inform them that all of their tests and quizzes will be kept for them in the folders. Although you should encourage students to take home test and quiz folders to show their parents or guardians, they are to bring them back to you.

Student folders can be kept in a file cabinet drawer or other specially marked area. Show the students where they're kept so they can pick them up and return them on their own.

At the end of the school year, you can return the folders to the students so they can use them in studying for final exams.

During the year, you can use the cumulative folder if you need to hold a parent conference dealing with a student's low grades. The folder provides visible evidence to the parent or guardian of the quality of the student's work.

1.18 BASING THE COURSE FINAL EXAM ON TESTS GIVEN DURING THE YEAR

When you return tests and quizzes, students often keep them just long enough to look at their grades, and then they stuff them in their notebooks or throw them in the trash basket. However, there is a technique that turns returned tests and quizzes into a learning tool. The value of the technique rests on the premise that the final exam for the course consists of questions taken solely from tests and quizzes given during the school year. After all, these previously asked questions represent the most important information that you thought students should know at the time. Why not use these questions for your final? In addition, this technique will allow students to study a manageable amount of material for the final.

Format for the Cover of Each Manila Folder

The format presented here is an example of one that has been used successfully. You might wish, however, to customize a format that is more convenient for your own needs.

MANILA FOLDER NUMBER SEQUENCE:

<u>1</u>

UNIT:

<u>Microbiology</u>

SPECIFIC LESSON TOPIC:

<u>Gram staining</u>

ITEMS IN FOLDER:	PURPOSE OF ITEM:
1. Gram staining lab sheets	1. Distribute to students for lab activity.
2. Notes introducing concept of the Gram stain	2. Present to class prior to lab activity.
3. Text study guide on the Gram stain	3. Distribute to students for homework.

ANY IMPORTANT ADDITIONAL NOTES?

• *Lab activity tends to run quite long the first time students attempt a Gram stain. Next year, take students through the procedure step by step.*

• *Important: Bacterial cultures should not be over 24 hours old.*

The basic steps of the technique are as follows:

1. When handing back the first test or quiz of the year, give each student a manila folder. Have each student put his or her name on the folder along with the name and section of the course. (See 1.17: The Student Cumulative Folder.)
2. Explain to students that the final exam (and midterm, if you have one) is comprised of questions taken from all tests and quizzes given during the school year. Therefore, all of the tests and quizzes returned during the year become valuable study guides for the final exam.
3. Upon receiving each returned test or quiz, the task is to correct each incorrect answer by writing in the proper answer. (You may want to allow students to use texts, notes, other students, or you as sources of correct answers.)

1.19   MAKING RESEARCH PAPERS INTERESTING: AN AUDIOVISUAL APPROACH

Many students dread even the mention of term papers. They panic, pondering the number of pages, footnotes, sources, and direct quotes they'll need. You certainly have students who feel this way and suffer from term paper burnout.

Many teachers also suffer from term paper burnout. Indeed, it is all but impossible for a teacher to evaluate a student's term paper validly. To do this, a teacher must not only read the term paper, but check the validity of the footnotes, bibliography, and quotes, in addition to checking for proper paraphrasing and evidence of plagiarism. Multiply this task by 25 or 30 term papers and the task becomes insurmountable.

Nevertheless, term papers are very important, and an educated student should be conversant with the proper format and techniques for researching and writing one. It might be helpful if students could take a semester-long course in term paper preparation. In such a course, the teacher might work with a limited number of students. The entire course might be taught in the school library. The teacher could work closely with the students, helping them and evaluating their progress in a step-by-step manner. The students might research a single overall topic, each taking a separate subtopic. This certainly would be a much more valid way of dealing with the term paper.

For those teachers who would occasionally like to assign an alternative to the conventional term paper, an audiovisual approach to a research topic can be both novel and stimulating to teacher and student alike. This approach emphasizes the use of a variety of audiovisual materials as the student presents research findings orally to the rest of the class. It provides an opportunity for students to be creative and to achieve the self-confidence and positive feedback that can result from a good presentation. In addition, the audiovisual approach allows the entire class to benefit from learning about each student's research. Finally, the audiovisual approach is a multimedia event that can capture and maintain the class's attention.

The following sequence of tasks can be carried out in preparing for and presenting research results:

1. Hand out copies of the evaluation sheet (Worksheet 1.19–1). This worksheet acquaints students with the criteria that you will be using in evaluating their audiovisual presentations. It will act as a guide for students as they prepare their presentations. (Examine this evaluation sheet carefully and make any modifications to meet your own standards.)

Name: _____   Date: _____

Evaluation Sheet for Audiovisual Presentations

Presentation subject:

Student's name:

Grade earned:

Criteria Used for Evaluating the Presentation:

1. Did presentation last for at least 30 minutes?

2. Was there undue repetition of material presented?

3. Was material presented in a way that those who have no previous knowledge of the subject matter could understand?

4. Was material presented in a way that made it interesting and enjoyable to the class?

5. Was all subject matter supplemented with a variety of audiovisual aids?

6. Which of the following audiovisual aids were incorporated into the presentation?

filmstrip projector	overhead projector	opaque projector
movie projector	VCR tape	models
computer graphics	cassette interview	guest speaker
commercial and/or prepared handouts	charts	chalkboard

Other materials used:

7. Were all supplemental materials neatly and adequately prepared?

8. Was the presenter familiar with the operation of audiovisual equipment used?

9. Was the presenter familiar with the correct pronunciation of terms used?

10. Additional notes and/or comments:

2. Select the research topic. Each student should select a topic of personal interest. The personal interest factor is crucial in motivating the student to do a fine job. You might allow a broad range of topics or perhaps have a general theme such as genetics, plants, animals, and so on.

3. Set presentation dates. In setting dates for each student's presentation, you might wish to accommodate those who would like to be among the first presenters. These students often feel the pressure of "getting up in front of the class" a little bit more than the others and will appreciate being able to "get it over with."

4. Instruct students in using the library media center. You might wish to arrange for the school librarian and/or audiovisual specialist to come to your class and present a minilesson on the uses of available library resources and facilities, as well as novel ways of preparing audiovisual aids.

5. Perform initial library work. After all students have selected a topic, take the class to the library for a period or two. The tasks for each student to accomplish and submit to you for approval are as follows: (a) five (you might want to change this number) bibliography sources containing materials relevant to the topic; (b) an initial, general outline of the topic including how each facet of the topic is going to be presented using an audiovisual aid.

6. Set the turn-in date for the finalized topic outline. Establish a date on which all finalized outlines should be turned in. Emphasize that students' outlines should include all audiovisual aids that are going to be used.

7. Make the audiovisual presentation. It is not necessary for students to memorize the material for their presentations. Note cards can be used. However, it is necessary that students supplement their entire presentations with audiovisual aids. During the presentation, fill out the evaluation sheet and record on it the grade earned. (You might want students to take relevant notes on each presentation, and then give a follow-up test at a later date.)

1.20   THE ANATOMY AWARDS PROGRAM

Introduction

The following material is based on nearly 20 consecutive years of running The Anatomy Awards Program. It describes in detail how you can run your own program. Naturally, you will want to modify suggestions and procedures to satisfy your own unique requirements in your school system. The Anatomy Awards Program is an excellent and fun activity to close out the school year. Students participate on a voluntary basis, and design, construct, and eventually wear costumes depicting structures studied in class during the year. All students who participate will long remember this activity, which could become an annual affair for you and your classes.

Determining the Scope of the Program

The first concern is to decide on the scope of The Anatomy Awards Program. You might decide to limit the program to one class period with participants from all of your classes. This is probably the least time consuming for you in the long run. If this is your choice, you might have to limit the number of students participating to, perhaps, 15 or 20 depending on classroom size and supervision. In this case, accept the participants on a first come, first served basis. Keep a list of a few students who can replace the occasional student who drops out of the activity.

The alternative is to have an Anatomy Awards Program for each class.

Obtaining Permission for Having the Program

The first step in organizing The Anatomy Awards Program is to obtain permission from the school administration. Once permission is obtained, select a date for the program. The date could be approximately three or four weeks prior to the end of school. (Have the date approved by the administration, since there could be a final date after which field trips and other activities may not be scheduled.)

Program Announcement and Student Selection of Costumes

Once permission has been obtained and a date scheduled, distribute copies of Worksheet 1.20–1 to the students. The worksheet should be distributed at least three months in advance of the actual program date. This will allow ample time for costume design and construction. Allow students a day or two to return the bottom portion of the worksheet forms indicating that they wish to participate in the activity. Have students write in the deadline dates on the worksheet. The deadline date for return of the "release from class" forms (Worksheet 1.20–3) should be soon after the student has indicated his or her desire to participate in the program. The deadline date for bringing the completed costume to school should be at least a week and a half to two weeks before the date of the actual program. This will allow some extra time for the student to make any costume modifications, if necessary. (The rain date is optional, depending on whether you choose to have the students parade costumes in the community. This option is discussed further under the heading "The Expanded Version of the Anatomy Awards.")

Once a student has selected a structure to depict, that student should show you, for your approval, a picture or a detailed sketch that will serve as a guide for costume construction. Structures selected for depiction will probably run the gamut from hearts, eyes, and lungs, to flowers, specific animals, molds, and bacteria. Although there will not be a lack of structures suitable for costumes, some students will probably need some guidance in selecting.

Monitoring Costume Construction

From the time students sign up for The Anatomy Awards until the program occurs, your main task is to monitor costume construction, answering any questions and/or helping students obtain materials. Distribute Worksheet 1.20–2. Page 1 of this worksheet will help guide participants with costume construction. Pages 2, 3, and 4 show

Name: _____   Date: _____

An Introduction to The Anatomy Awards Program

Dear Student,

The Anatomy Awards Program is a strictly voluntary activity. If you wish to participate in this year's program, it will involve the following responsibilities on your part:

1. Selecting a structure studied in class
2. Designing and constructing a costume depicting the structure
3. Wearing the costume as a part of The Anatomy Awards Program.

Costumes can be constructed out of any material that you choose. Materials can range from cardboard and burlap to chicken wire covered with papier-mâché. The two major requirements for the costume are (a) it must accurately represent the structure it is depicting, and (b) it must cover the entire body.

On the day of The Anatomy Awards Program, judges will be present to select the best costumes. One costume will be judged as The Anatomy Award Grand Prize Winner. Other costumes will be judged as runner-up winners. Prizes will be awarded to all of the winners, and all participants will receive a Certificate of Participation.

The following deadline dates are of importance:

_____: For teacher approval of your structure for costume design. (Include a picture or detailed sketch that you are going to follow in designing and constructing your costume.)

_____: For release-from-class forms to be returned to your teacher.

_____: For bringing your completed costume to school.

_____: The date and class period of this year's Anatomy Awards Program.

_____: The rain date (if necessary).

If you wish to participate in The Anatomy Awards Program, please fill out the form below, detach it along the dotted line, and give it to your teacher. Participation is on a first come, first served basis.

YES!!!! I want to participate in this year's Anatomy Awards Program.

Student Name:

Name and Section of Course:

Costume Idea (if you already have one):

WORKSHEET 1.20–2

The Anatomy Awards Costume Construction Guide

The following are some guidelines and information that will help you as you begin construction of your Anatomy Awards Costume.

- It is essential to begin costume construction as soon as possible. Unexpected, time-consuming problems have a way of happening, and this can be a major problem if costume construction is put off until the last minute.

- Costumes are often quite large. Be certain that costumes will be able to fit through the door of the classroom. For costumes that are going to present a problem in this regard, designing the costume in two sections may solve the problem.

- Chicken wire (a lightweight wire with a 1- or 2-inch mesh) can be obtained at a hardware store. It is inexpensive and can easily be bent in shaping the desired structure for your costume.

- Cardboard and cardboard boxes are fine materials for costume construction. Appliance stores are good sources for large boxes that once contained refrigerators, stoves, and so on. Often, appliance stores are happy to get rid of them.

- Burlap and canvas are also good materials for costume construction.

- Do not create a costume that is too heavy; remember, you are going to be wearing it.

- Papier-mâché is excellent for covering basic shapes formed from chicken wire or cardboard. To papier-mâché, you first prepare a paste mixture of flour and water that is approximately the same consistency as pancake mix. Guard against a mixture that is either too watery or too thick. Then tear newspaper sheets into sections running the length of the paper and measuring approximately 5 inches to 6 inches wide. Moisten the paper sections with the paste mixture. This can be achieved by dipping the sections into the paste, allowing any excess paste to run off. Cover the basic shape formed by the chicken wire or the cardboard with one or two layers of the moistened newspaper. (You might have to cover it one half at a time, allowing it to dry before covering the other half.) Allow the layer(s) to dry thoroughly. This usually takes two to three days. Then add a couple of more layers, and allow the drying time. If the structure needs more support, add more layers. (*Note:* Working with papier-mâché can be very messy. Ask your parents/guardians where the best location would be for you to do your work.) Painting the dried costume is easy.

- *Important Safety Caution:* When working with material such as chicken wire, be certain that all sharp pieces of wire are removed with a wire cutter. If not, you can be seriously injured when putting on, wearing, or taking off your costume.

This page and the ones that follow show pictures of actual costumes made by Anatomy Awards participants.

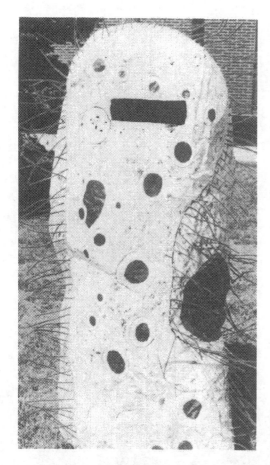

This costume was constructed out of papier-mâché. The inner framework is chicken wire. The cilia around the outside is made of pipe cleaners. A rectangle for vision has been cut out at the top. Inside the costume are two wooden grip handles for the hands to hold onto when wearing.

1. A Paramecium

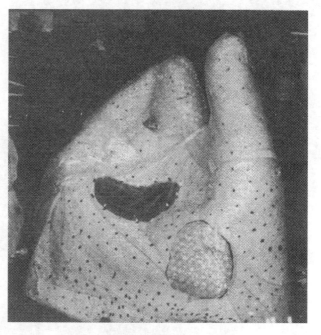

2. An Amoeba

Constructed of papier-mâché over a chicken wire framework, this costume has been painted white and is outlined with a set of blinking lights (holiday type) that draw attention to various structures. The lights operate off a small power pack. The cut-away area for vision is on the opposite side of the costume. The costume has been covered with protective plastic.

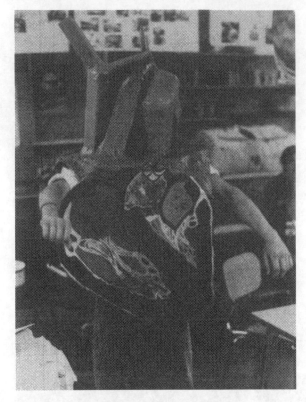

This costume is made totally of cardboard and some tape to attach the vessels. It shows internal structures of an auricle and a ventricle. Veins were painted blue and arteries red. Arm holes were cut for ease of wearing.

3. The Human Heart

4. A DNA Molecule

Over an inner framework of chicken wire, burlap was added. The sugar, phosphate, and nitrogen base units were color coded. A half of a DNA molecule is at the right of the picture, waiting to move into position when the full molecule replicates.

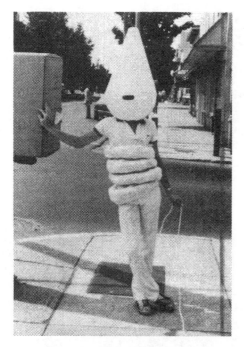

This costume shows the head, body portion, and tail of a spermatozoon. The head, with visual eyehole, is papier-mâché over oak-tag. The body is a stuffed cloth "tube." The tail is made of clothesline.

5. A Spermatozoon

6. The Buccal (mouth) Cavity

This incredible costume was designed as a float. The lips were made of cardboard, the tongue of cloth, and the teeth were cut out of pieces of plastic foam material. Sheets help cover the shopping cart (borrowed with permission of a local food market) on which the costume rides. Three students, hidden inside the costume, open and close the mouth as the float moves along.

Release-from-Class Form for Participating in The Anatomy Awards Program

Dear Teacher:

Please sign below if you will excuse this student from your class on_____ (rain date, if any: _____) to participate in a special end-of-year science program called The Anatomy Awards.

If I can provide you with any further information, please let me know.

Thank you.

Sincerely,

Student Name _____

Teacher Signature _____

Dear Teacher:

Please sign below if you will excuse this student from your class on_____ (rain date, if any: _____) to participate in a special end-of-year science program called The Anatomy Awards.

If I can provide you with any further information, please let me know.

Thank you.

Sincerely,

Student Name _____

Teacher Signature _____

Dear Teacher:

Please sign below if you will excuse this student from your class on_____ (rain date, if any: _____) to participate in a special end-of-year science program called The Anatomy Awards.

If I can provide you with any further information, please let me know.

Thank you.

Sincerely,

Student Name _____

Teacher Signature _____

Anatomy Award costumes that students made. (Some copiers will reproduce these pictures; others will not.)

Storing Costumes at the School:

You will have to arrange for an area where you will keep the costumes once students start bringing them to school. The location should be one that is safe and that keeps the costumes out of sight until the day of the program.

Locating the Area in Which the Program Will Be Held:

The classroom is an ideal area in which to hold the program. If you have a planning period when your classroom is empty, you could hold the program during that period. If not, you could hold the program during one of your regular class periods. In the latter case, students who are not participating could be provided with work and sent to the school library. (Obtain permission in advance from the librarian.)

Obtaining Permission for Students to Participate in the Program:

Participating students will have to be excused from their classes for the period, but that should pose no major problem. Worksheet 1.20–3 provides permission slips that can be used for this purpose. Permission slips should be given to students soon after they have volunteered to participate in the program, and they should be returned to you as soon as possible.

Photographing The Anatomy Awards Program:

If you wish to record The Anatomy Awards Program on slides, color prints, and/or videotape, select two or three students who are competent photographers. Capturing The Anatomy Awards on videotape and 35-mm color slides will allow you to show them to next year's students as motivation for participation. The photographs are fine for classroom posters depicting The Anatomy Awards.

Inviting Parents and Guardians to the Classroom Program:

Most parents and guardians become actively involved in helping students design and construct their costumes. If you invite them to the classroom program, several will probably show up. This is excellent public relations, and you can be certain that those who attend will spread word about "one of most unusual and innovative programs ever seen."

Selecting the Anatomy Award Costume Judges:

Potential sources of judges include past student graduates (who usually are more than delighted to return to the school for this program); school nurses and/or school health care workers; and other teachers who might be free during the period of the program. Selecting an uneven number of judges (for example, five) allows for one judge to break a voting tie.

After the first year of The Anatomy Awards Program, judges can be selected from past participants.

Selecting the Master of Ceremonies:

You will need a student to act as master of ceremonies (MC) for the program. Prior to the program, review with the MC the sequence of program events. Also have participating students write on 3 × 5 cards the descriptions of their costumes that they want the MC to read during the judging.

Arranging for Program Refreshments:

You can serve refreshments as a conclusion to the program. An easy way to obtain all of the food needed is to ask each participant to bring in a food item, such as pizza, hot dogs, rolls, potato salad, cole slaw, potato chips, sodas, eating utensils (such as paper plates and plastic knives, forks, and spoons), napkins, condiments, desserts (such as cakes and cookies), and so on. You will have to estimate the quantity of each item needed based on the number of participants. (*Note*: Food items will vary depending on whether you have a way of cooking and/or heating foods.)

Items that will not perish should be brought into school a day in advance. If you have access to a refrigerator, perishable items can also be brought in in advance. This assures that everything is ready and that a student has not forgotten to bring in a food item at the last minute.

To simplify matters, Worksheet 1.20–4 provides slips that can be used in assigning food items to participants.

Selecting Prizes to Be Awarded for the Best Costumes:

You will have to decide what types of prizes will be awarded. If you feel that monetary prizes are suitable, your science department or school system might provide the money. Textbook and/or lab manuals that are no longer used in the classroom (yet are often up to date in terms of content) make ideal prizes. Students are usually willing to donate items, such as a rock star poster. In addition to prizes, all participants should receive a Certificate of Participation (Worksheet 1.20–5).

Notifying the Media:

An excellent source of public relations is to invite a local newspaper and/or television station to cover The Anatomy Awards Program. They are usually eager to oblige, because this type of human interest event appeals to the media. Send the media a letter explaining The Anatomy Awards Program. Follow up with a personal phone call.

Forms for Assigning Food
to Be Brought in for the Program

Student Name: Food Item: Quantity:	Student Name: Food Item: Quantity:
Student Name: Food Item: Quantity:	Student Name: Food Item: Quantity:
Student Name: Food Item: Quantity:	Student Name: Food Item: Quantity:
Student Name: Food Item: Quantity:	Student Name: Food Item: Quantity:
Student Name: Food Item: Quantity:	Student Name: Food Item: Quantity:
Student Name: Food Item: Quantity:	Student Name: Food Item: Quantity:
Student Name: Food Item: Quantity:	Student Name: Food Item: Quantity:

Anatomy Awards Certificate of Participation

The Anatomy Awards

This is to certify that

_____

has been awarded the Official

Certificate of Participation

for the year _____

_____
Teacher

ANATOMY·AWARD·SEAL

Arranging the Classroom on the Day of the Program:

The classroom can be arranged simply by moving desks and chairs against the walls. A few desks should be grouped together to serve as the refreshment area. Food items and utensils can be arranged on the desks for use at the conclusion of the program.

A few desks and chairs should be arranged in a row to provide seating for the judges. Copies of Worksheet 1.20–6 can be used as nameplates for the judges and should be taped on the desk in front of each judge. Copies of Worksheet 1.20–7, which explains the procedure for judging costumes, should be placed on the judges' desks.

A large masking tape "X" on the floor can denote where participants should stand during the judging.

Two or three trash cans should be located about the room. This will facilitate clean-up at the end of the program.

If you have access to a small portable lectern with a microphone, you can place it at the front of the classroom for use by the MC. It is used solely to enhance ("ham up") the program. At the lectern, place the materials that the MC will need (namely, the program of events and the 3 × 5 cards containing the costume descriptions).

You might wish to have a cassette player with a tape of fanfares to be played prior to the judging of each costume and/or the awarding of prizes.

Outline of The Anatomy Awards Program:

A program outline similar to that shown in Worksheet 1.20–8 should be presented to all participants. (Note that the program refers to the presentation of a monetary donation to a health organization. This possibility is discussed in the next section.)

The Expanded Version of The Anatomy Awards:

The Anatomy Awards Program can be expanded to include a community activity such as collecting monies for donation to a health organization. Students can be involved in selecting the recipient organization. For example, it could be an organization involved in research dealing with heart disease, lung disease, or cancer. Or it could be a local nursing home or association for the blind. Notify the recipient organization and ask it to send a representative to the classroom program.

This expanded version can make The Anatomy Awards Program truly outstanding, and it is definitely worth the extra effort. There are several modes of collecting the money:

- Do it on a schoolwide basis, through homeroom. (Get permission from the administration.)

- Obtain a community booth at a local mall for use on a Saturday afternoon. You and three or four students, wearing their costumes, will staff the booth. (Get permission from the mall, the administration, and the parents or guardians of the involved students. Also be absolutely certain that you and the students are covered by appropriate accident and liability insurance through the school district.)
- Visit local community business establishments with students, in costume. (Get permission from the administration, the community, and the parents of the involved students. Again, school district accident and liability insurance is mandatory.)

Nameplates for the Judges

THE ANATOMY AWARDS PROGRAM PROUDLY ANNOUNCES THE
NAME OF:

_____

PARTICIPATING AS A COSTUME JUDGE

Costume Rating Sheet for the Judges

Instructions:

1. One by one each participant will model his or her costume. The MC will read a description of each costume as previously prepared by the participant.
2. Each costume should be rated on a scale from 1 to 10, with 10 being the best.
3. As each costume is judged, please fill in the information below. (Under the heading "Pertinent Information," you might want to note any outstanding attribute[s] of the costume.)
4. At the completion of the judging, compare ratings given by each judge as well as any information recorded in the "Pertinent Information" column. During this discussion, a judge may change his or her rating given to a costume.
5. Finally, add up the rating scores for each costume. The top score will be considered the grand prize winner. The next three highest scores will be considered the runners up.

COSTUME NAME	PERTINENT INFORMATION	RATING

Sample Classroom Program for the Anatomy Awards

1. Participants put on their costumes in preparation for the judging.
2. Opening of the program by the teacher:
   a. Welcoming everyone to The Anatomy Awards Program
   b. Introducing the individual(s) representing the health organization recipient of a monetary donation
   c. Introducing any parents present
   d. Introducing any media personnel present
   e. Introducing the MC
3. Continuation of the program by the MC:
   a. Welcoming and introducing the judges
   b. Introducing the student photographer who will be taking pictures of each costume during judging
   c. Reading the costume description for each costume as it is being judged
   d. Announcing the winners and presenting to them the prizes
   e. Distributing the Certificates of Participation
   f. Introducing the student who will present the monetary donation to the health organization recipient
4. Student presentation of the monetary donation
5. Conclusion of the program by the teacher:
   a. Thanking once again all who participated in any way in making The Anatomy Awards Program a success
   b. Announcing that refreshments are now being served

Autographs:

The Grand Prize Winner: _____

The Three Runners Up: _____

_____

_____

The Judges:

_____   _____

_____   _____

_____   _____

_____   _____

The MC: _____

The Photographer: _____

The Teacher: _____

1.21 LESSON PLANNING

Goals are vital in classroom teaching. In addition, we must be able to defend on strong educational ground the course content that we teach, the methodology that we use in teaching it, and why we have set the goals that we have. Too often we teach the same materials over and over, year after year, without considering whether it is time to update or overhaul our teaching content and methodology.

The veteran teacher can probably recall those days of student teaching when in-depth daily lesson plans had to be prepared replete with short- and long-term goals. It is easy, over years of teaching, to relegate formal lesson planning to the back burner. Lesson plans, over time, can become established in your head and thus are not articulated on paper. However, over years of teaching you can lose sight of whether your lessons are achieving the goals that they were originally designed to achieve.

So, whether you enjoy it or not, classroom lesson plans must be updated periodically. Using Worksheets 1.21–1 through 1.21–4 will make this task much easier.

Worksheet 1.21–1 lets you itemize the subject matter covered, as well as the general goals sought for each unit of material covered. Worksheet 1.21–2 is a completed sample.

Worksheet 1.21–3 lets you focus on daily lesson plans complete with more specific goals, teaching techniques used to meet them, and equipment needed. Worksheet 1.21–4 is a completed sample.

1.22 STUDENT NOTE TAKING AND ORGANIZATION OF MATERIAL

A major obstacle to learning and understanding subject matter is the way in which students organize their class notes prior to studying. Often the materials discussed in class are not presented in the order that will maximize student learning during later study sessions. For example, the structure and function of the ear might be presented to the class using a lecture format. Then a filmstrip on the ear might be used for learning reinforcement while additional notes on the ear are presented. Finally, an article on the ear along with a study guide for further notes might be given to the class. When it comes time for the student to study the material for an exam, the material presented in class should be arranged in a logical format to facilitate memory. Unfortunately, there are far too many students who do not take the time needed to put their notes in order. Therefore the material being studied is disjointed.

To give students practice in organizing notes, pass out copies of Worksheet 1.22–1. Instruct students to organize the notes on another sheet of paper. After the task is accomplished, project selected papers on the classroom screen using an opaque projector and conduct a class discussion on how well the material was organized. Worksheet 1.22–2 shows one example of how the material can be organized in a logical format.

Repeat the preceding exercise using Worksheet 1.22–3. Worksheet 1.22–4 shows one example of how this material can be organized.

Lesson Plan Overview for a Unit of Material

UNIT NAME:

APPROXIMATE # OF WEEKS TO COMPLETE:

TOPICS	GOALS
1.	
2.	
3.	
4.	
5.	
6.	
7.	
8.	
9.	
10.	
11.	

Sample Lesson Plan Overview for a Unit of Material

UNIT NAME: The skeletal system

APPROXIMATE # OF WEEKS TO COMPLETE: 3

TOPICS	GOALS
1. the human skeleton	a. names of the bones b. names of the sutures
2. the typical long bone	a. parts of the bone b. functions of the parts
3. microscopic structure of bone	a. names of parts b. functions of parts
4. bone projections and processes	a. terms used for b. definitions of c. names of d. examples of
5. individual bones of the skeleton	a. identification of b. landmarks on
6. the vertebral column	a. sections of b. functions of c. clinical problems of
7. types of joints	a. names of b. definitions of c. examples of
8. typical joint structure	a. parts of b. functions of c. clinical problems of
9. joint movements	a. names of b. definitions of c. examples of
10. divisions of skeleton	a. names of b. descriptions of c. bones associated with
11. diseases of skeletal system	a. names of b. causes of c. descriptions of

A Format for Daily Lesson Plans

UNIT NAME: _____

SPECIFIC TOPIC: _____

LESSON #: _____

GOALS	TEACHING TECHNIQUES	EQUIPMENT NEEDED

— —

NOTES TO SELF:

A Sample Format for Daily Lesson Plans

UNIT NAME: _____the skeletal system_____
SPECIFIC TOPIC: __bones of the skeleton__
LESSON #: __1__

GOALS	TEACHING TECHNIQUES	EQUIPMENT NEEDED
1. To determine the extent of student knowledge of the bones of the human skeleton	a. Sponge activity: labeling a blank diagram of the skeleton and blank diagram of the hand and wrist	• Blank diagrams of skeletal system
2. To learn the names of the bones of the human skeleton	a. Videotape: "The Skeletal System" (15 min.)	• Video from media center • VCR monitor
	b. Students label diagram of skeleton using text as source	• Skeleton
	c. Students locate bones on actual skeleton	• Skeleton transparency
	d. Show overhead transparency of skeleton on screen and have volunteers point out and name bones	• Overhead projector
	e. Announce a quiz tomorrow requiring the labeling of a diagram of the skeleton	• Blank diagrams of skeleton
3. Time permitting, begin learning the names of the bones of the hand and wrist	a. Students label diagrams of the hand and wrist using transparency on screen as a guide	• Blank diagrams of hand and wrist
	b. Students locate bones on actual hand and wrist of skeleton	• Hand and wrist models

— —

NOTES TO SELF:

A more in-depth video on the skeletal system is needed.
Keep a spare overhead projector bulb on hand.
Carpals coming loose on one of the models.

WORKSHEET 1.22–1

Note Organization Exercise 1

The blind spot does not contain rods or cones.

Rods are not sensitive to color.

The middle layer of the eye is called the choroid.

The eye has two main cavities, the anterior and the posterior.

The function of the retina is to convert light waves to nerve impulses.

Myopia is the term for nearsightedness.

The function of the sclera of the eye is that of protection.

The fovea is the only area on the retina for sharp focus of light waves.

The lens separates the anterior and the posterior cavities of the eye.

The anterior cavity of the eye contains aqueous humor.

Two types of visual problems are nearsightedness and farsightedness.

The function of the choroid is to prevent light reflection inside the eye.

The innermost layer of the eye is the retina.

The lens is held in place by structures called suspensory ligaments.

Hyperopia is caused by an eyeball that tends to be too short.

Cones are sensitive to color.

The posterior cavity of the eye contains vitreous gel.

The function of the lens is to help focus light waves.

Myopia is caused by an eyeball that tends to be too long.

The retina is the photosensitive layer of the eye.

The outer layer of the eye is the sclera.

Another name for farsightedness is hyperopia.

Rods are primarily used for night vision.

The iris is located in the anterior cavity of the eye.

Myopia can be corrected with concave lenses.

The retina contains photosensitive nerve cells called rods and cones.

The eye has three layers.

Hyperopia can be corrected with convex lenses.

Two areas of significance on the retina are the fovea and blind spot.

The iris is the colored part of the eye.

Cones are of three types: red sensitive, green sensitive, and blue sensitive.

Cones are used primarily for daylight vision.

WORKSHEET 1.22–2

One Example of Logically Organized Notes

The eye has three layers.

The outer layer of the eye is the sclera.

The function of the sclera of the eye is that of protection.

The middle layer of the eye is called the choroid.

The function of the choroid is to prevent light reflection inside the eye.

The innermost layer of the eye is the retina.

The function of the retina is to convert light waves into nerve impulses.

The retina is the photosensitive layer of the eye.

The retina contains photosensitive nerve cells called rods and cones.

Rods are not sensitive to color.

Rods are used primarily for night vision.

Cones are sensitive to color.

Cones are of three types: red sensitive, green sensitive, and blue sensitive.

Cones are primarily used for daylight vision.

Two areas of significance on the retina are the fovea and blind spot.

The blind spot does not contain rods or cones.

The fovea is the only area on the retina for sharp focus of light rays.

The eye has two main cavities, the anterior and the posterior.

The posterior cavity of the eye contains vitreous gel.

The anterior cavity of the eye contains aqueous humor.

The lens separates the anterior and posterior cavities of the eye.

The function of the lens is to help focus light waves.

The lens is held in place by structures called suspensory ligaments.

The iris is located in the anterior cavity of the eye.

The iris is the colored part of the eye.

Two types of visual problems are nearsightedness and farsightedness.

Myopia is the term for nearsightedness.

Myopia is caused by an eyeball that tends to be too long.

Myopia can be corrected with concave lenses.

Another name for farsightedness is hyperopia.

Hyperopia is caused by an eyeball that tends to be too short.

Hyperopia is corrected with convex lenses.

WORKSHEET 1.22–3

Note Organization Exercise 2

The temporal lobe is involved in hearing.

The CNS consists of the brain and the spinal cord.

Sensory neurons carry nerve impulses from body receptors to the brain and spinal cord.

The midbrain deals with pupillary reflexes.

Peripheral nerves are made up of sensory and motor neurons.

The two main divisions of the nervous system are the central and peripheral.

Of the 43 pairs of peripheral nerves, 12 pairs are cranial nerves.

The names of the visible lobes of cerebrum are frontal, parietal, temporal, and occipital.

The peripheral nervous system is abbreviated PNS.

The cerebellum deals with muscle coordination.

Motor neurons carry nerve impulses from the brain and spinal cord to muscles and glands.

The occipital lobe is involved in sight.

There are 43 pairs of peripheral nerves that connect the brain and spinal cord to all parts of the body.

The central nervous system is abbreviated the CNS.

The medulla oblongata deals with reflexes such as sneezing.

The parietal lobe is involved in taste.

Of the 43 pairs of peripheral nerves, 31 pairs are spinal nerves.

Three sections of the brain are the cerebrum, cerebellum, and midbrain.

The cerebrum is divided into four visible lobes.

The PNS consists of pairs of nerves that connect the brain and spinal cord to all parts of the body.

The frontal lobe is involved in thinking and planning ahead.

Three parts of the brain stem are the midbrain, medulla oblongata, and pons.

The pons deals with regulating respiration.

One Example of Logically Organized Notes

The two main divisions of the nervous system are the central and peripheral.

The peripheral nervous system is abbreviated PNS.

The central nervous system is abbreviated CNS.

The CNS consists of the brain and the spinal cord.

Three sections of the brain are the cerebrum, cerebellum, and midbrain.

The cerebrum is divided into four visible lobes.

The names of the visible lobes of the cerebrum are frontal, parietal, temporal, and occipital.

The frontal lobe is involved in thinking and planning ahead.

The occipital lobe is involved in sight.

The temporal lobe is involved in hearing.

The parietal lobe is involved in taste.

The cerebellum deals with muscle coordination.

Three parts of the brain stem are the midbrain, medulla oblongata, and pons.

The midbrain deals with pupillary reflexes.

The pons deals with regulating respiration.

The medulla oblongata deals with reflexes such as sneezing.

The PNS consists of pairs of nerves that connect the brain and spinal cord to all parts of the body.

There are 43 pairs of peripheral nerves that connect the brain and spinal cord to all parts of the body.

Of the 43 pairs of peripheral nerves, 12 pairs are cranial nerves.

Of the 43 pairs of peripheral nerves, 31 pairs are spinal nerves.

Peripheral nerves are made up of sensory and motor neurons.

Motor neurons carry nerve impulses from the brain and spinal cord to muscles and glands.

Sensory neurons carry nerve impulses from body receptors to the brain and spinal cord.

Preparing several of your own note organization exercises for the students to work on from time to time can go a long way toward improving students' note organization abilities.

1.23 DEVELOPING EFFECTIVE LISTENING SKILLS IN STUDENTS

A disconcerting problem can arise when you are giving classroom notes or instructions: Students constantly ask you to repeat information just given. From time to time, of course, a student has a legitimate right to ask you to repeat. Perhaps you were giving notes or instructions at a fast pace. However, when giving notes and instructions becomes a chore because you constantly have to repeat yourself, it is time to take action.

This problem can be eliminated by encouraging students to do the best they can in taking notes. Tell them that if they miss any notes, they can see you after class and you will help them fill in any missing notes. When giving instructions (for example, just prior to a laboratory activity), inform students that if anyone has questions you will be available to answer them as the activity begins.

Even if you eliminate the aforementioned problem, you have not dealt with the cause of the problem. The cause could be that students lack effective listening skills. Exercises designed to improve students' listening skills are sadly lacking in many classroom curricula. Effective listening is a topic that should be dealt with on an ongoing basis, thus enabling students to become better listeners.

Worksheets 1.23–1 through 1.23–8 are designed as beginning exercises to acquaint students with some facts about learning and allow them to improve their listening skills.

Although the listening skill activities presented here are rather simple, some students will nonetheless have difficulty in achieving 100% accuracy. Develop several other listening skills exercises and use them throughout the school year. They are fun to develop, and students enjoy doing them.

Listening skills activities should challenge students to deal with increasingly larger pieces of instructional information at one time.

1.24 BEHAVIORAL OBJECTIVES

It is essential that any course syllabus contain a set of clearly defined goals that provide specific direction for

1. You in your teaching activities
2. Students in their learning activities
3. The program evaluator in his or her evaluating activities.

Such a set of clearly defined goals can be formulated in terms of behavioral objectives.

During the 1960s and 1970s, a great deal was written about the value of developing course objectives using a behavioral objective format. For example, Robert F. Mager (*Preparing Instructional Objectives,* Palo Alto: Fearon Publishers, Inc., 1962, p. 3)

Listening Techniques Notes to Be Given to Students

Teacher Instructions: Distribute copies of Worksheet 1.23–2 to students to use as a format for note taking as you give them the following information.

Effective listening: is not easy.

Effective listening requires: conscious and constant effort on the part of the listener.

Effective listening is actually a: physiological process.

Although not noticed by the listener, effective listening causes:

1. The heart rate to increase.
2. The palms to sweat.

Examples of situations in which the listener appears to be listening but really isn't:

1. The listener appears to be paying attention, but is actually daydreaming, not hearing a word that is being said.

2. The listener appears to be paying attention, appears to be not daydreaming, and is actually hearing the words being said. However, the words are being heard as just that, words! They are "going in one ear and out the other."

Effective listening is going on when:

1. The listener is listening with an active mind.
2. The listener is not only hearing the words being said, but is also processing the information.

Processing information means:

As the information is being received by the listener, it is being reinforced and given meaning by the listener, so that it is more likely that it will be remembered.

Techniques of processing information:

1. Writing down the information in the form of notes
2. Mentally repeating the information word for word
3. Paraphrasing the information
4. Converting the information into visual form in your "mind's eye," such as on a mental TV screen

Becoming an effective listener results in: becoming a more effective learner with an improved memory.

Name: _____ **Date:** _____

WORKSHEET 1.23–2

Student Format Sheet for Note Taking

Student Instructions: As your teacher gives you notes on listening, record them using the following format:

Effective listening:

Effective listening requires:

Effective listening is actually a:

Although not noticed by the listener, effective listening causes:

1.

2.

Examples of situations in which the listener appears to be listening, but really isn't:

1.

2.

Effective listening is going on when:

1.

2.

Processing information means:

Techniques of processing information:

1.

2.

3.

4.

Becoming an effective listener results in:

Teacher Format for
Map Drawing Listening Exercise

Teacher Instructions:

a. Distribute copies of Worksheet 1.23-4 to students to use as a format for drawing a map.
b. Go over the instructions on the student worksheet with students.
c. Make sure students realize that no instruction will be repeated.
d. Read each of the steps below slowly and clearly, leaving sufficient time between steps for students to carry out each task. (It will help if you carry out this exercise yourself prior to giving it to students. This will allow you to get a feel for how much time to leave between steps.)
e. Following this exercise, select several student maps for projecting on the classroom screen using an opaque projector. A completed sample is shown in Worksheet 1.23–5.

Steps:

1. Go north to second traffic light and turn west.

2. Proceed west until you come to the medical center on your left.

3. Turn left just after passing the medical center.

4. Go straight and turn right just after crossing railroad tracks.

5. Continue west and turn north at the third intersection.

6. Go north until you come to Ridgeway Park.

7. Follow the road around Ridgeway Park and continue east until you cross a bridge.

8. Just after crossing bridge, turn left.

9. Go straight for a ways and turn left on Pine Street.

10. Continue on Pine Street, turning right at the fourth intersection.

11. Pass a church to your west and then gas station on your left.

12. Turn right just after gas station.

13. Your destination is a parking lot on the south side of the highway.

WORKSHEET 1.23-4

Student Format for
Map Drawing Listening Exercise

Student Instructions:

a. Your teacher will orally give you step-by-step instructions for drawing a map to use in driving a car to a particular location.
b. As you listen to each step read to you, draw it in the form of a map.
c. Each step in the instructions will be read only once, so listen carefully.
d. Map symbols to incorporate into your map are provided at the bottom of this worksheet. (Also note direction indicator.)
e. Begin your map at START.

START

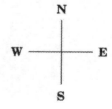

RP = Ridgeway Park	Pine St. = Pine Street	MC = medical center
P-LOT = parking lot	GAS = gas station	▰▰▰ = railroad tracks
C = church	▬▬▬ = bridge	Direction Indicator
+ = intersection	0 = traffic light	

WORKSHEET 1.23-5

Completed Sample for
Map Drawing Listening Exercise

Student Instructions:

a. Your teacher will orally give you step-by-step instructions for drawing a map to use in driving a car to a particular location.
b. As you listen to each step read to you, draw it in the form of a map.
c. Each step in the instructions will be read only once, so listen carefully.
d. Map symbols to incorporate into your map are provided at the bottom of this worksheet. (Also note direction indicator.)
e. Begin your map at START.

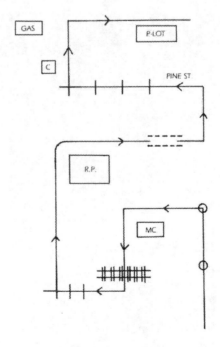

RP = Ridgeway Park   Pine St. = Pine Street   MC = medical center

P-LOT = parking lot   GAS = gas station   ‖‖‖ = railroad tracks

C = church   ▀▀▀ = bridge   Direction Indicator

+ = intersection   0 = traffic light

Teacher Format for Geometric Drawing
Listening Exercise

Teacher Instructions:

a. Distribute copies of Worksheet 1.23–7, along with rulers, to students.
b. Go over the instructions on the student worksheet with students.
c. Make sure that students realize that no instruction will be repeated.
d. Read each of the steps below slowly and clearly, leaving sufficient time between steps for students to carry out each task.
e. Following this exercise, select several student maps for projecting on the classroom screen using an opaque projector.

Steps:

1. Draw a 1-inch square in the approximate center of the worksheet.
2. Beginning at the upper left-hand corner of the square, draw a 2-inch line in the direction of northwest.
3. One inch directly below the square, and centered with the square, draw a circle approximately the size of a quarter.
4. Put a dot in the center of the circle.
5. From the tip of the 2-inch line that you drew to the northwest, draw a 1-inch line straight east.
6. From the dot in the center of the circle, draw a straight line to each of the bottom corners of the square.
7. Draw a straight line from the open end of the 1-inch line that you drew going east, to the upper right-hand corner of the square.
8. From the upper northwest tip of your figure, draw a 1-inch line straight south.
9. From the open end of the line that you just drew, draw a line straight east stopping at the first line with which it intersects.
10. Shade in the inverted triangle that has its tip in the center of the circle.
11. From the center of the circle, draw a 2-inch line straight east.
12. Draw a dotted line extending 2 inches directly north and 1 inch directly south of the open end of the line you just drew.
13. From the open end of the 2-inch dotted line, extend a straight dotted line until it meets the northwest tip of the figure.

Name: _____ Date: _____

Student Format for
Geometric Drawing Listening Exercise

Student Instructions:

a. Your teacher will orally give you step-by-step instructions for drawing a geometric figure.
b. As you listen to each step read to you, draw it on this worksheet.
c. Each step in the instructions will be read only once, so listen carefully.
d. Note the direction indicator at the bottom of this worksheet.

Direction Indicator: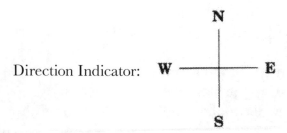

Name: _____ **Date:** _____

Completed Sample for
Geometric Drawing Listening Exercise

Student Instructions:

a. Your teacher will orally give you step-by-step instructions for drawing a geometric figure.

b. As you listen to each step read to you, draw it on this worksheet.

c. Each step in the instructions will be read only once, so listen carefully.

d. Note the direction indicator at the bottom of this worksheet.

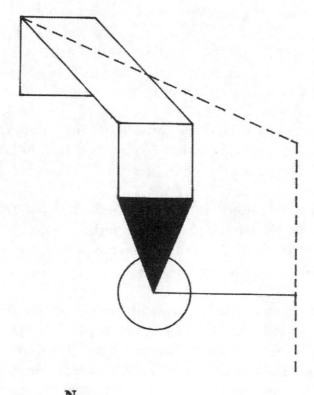

Direction Indicator: W ——|—— E   (N top, S bottom)

wrote: "When clearly defined goals are lacking, it is impossible to evaluate a course or program efficiently and there is no sound basis for selecting appropriate materials, content, or instructional methods." Unfortunately, the behavioral objective approach to education proved, in retrospect, to be rather short lived. Even so, what W. James Popham ("Behavioral Objectives and Teaching Skills," *Forum*, 8:4–7, October, 1969, p. 4) wrote over two decades ago is as true and vital today as it was then:

> What do I want students to become? What kinds of modifications will I try to effect in them? Having asked that question, then I can make better decisions about what I do and what they do. But to ask that question properly requires that one explicate his goal, talk about his intentions in terms of changes in learners.

In other words, to construct a truly meaningful and relevant course syllabus, you must first decide what student behavior changes are desirable and then articulate the syllabus in terms of the appropriate behavioral objectives.

The ideal format used in designing a behavioral objective assures that each objective includes the following three pieces of information:

1. The specific behavior that the student is to learn
2. The specific way in which the student will be tested to ascertain learning
3. Specifically, what will constitute evidence of success in learning.

Assume that, as part of a course syllabus, it is important for students to be able to ignite a Bunsen burner properly before participating in laboratory activities. A behavioral objective designed for such learning might be written as follows:

> The student will evidence ability to ignite a Bunsen burner properly by carrying out in sequence the following events: lighting a match, holding the lighted match over the barrow of the burner, turning on the gas jet, and successfully igniting the burner without an undue burst of flame.

The preceding objective contains the three pieces of information required for an objective to be behavioral. One, what is the specific behavior that the student is to learn? (Answer: To ignite a Bunsen burner properly) Two, what is the specific way in which the student will be tested to ascertain learning? (Answer: By actually igniting a Bunsen burner) Three, what specifically constitutes evidence of success in learning? (Answer: Igniting the burner using the proper sequence of events, and without an undue burst of flame)

As another example, assume that learning bones of the body is part of a course curriculum. Two behavioral objectives designed for a portion of this unit could be as follows:

- When given a diagram of the bones of the human wrist, the student will be able to label correctly, with correct spelling, all eight carpal bones.
- When shown the actual skeleton of the human wrist, the student will be able to point out, correctly identify, and correctly pronounce the names of all eight carpal bones.

The behavioral objective provides a viable means of accountability in the classroom, because it clearly states the objectives of your classroom endeavors. Be sure to state objectives as behavioral objectives when preparing a course syllabus.

SECTION 2

Success-Directed Learning in the Classroom

Section 2 provides step-by-step techniques, including nine reproducibles, for initiating a system that holds students accountable for their own learning. These techniques focus on teamwork between you, the student, and the parent or guardian in assuring student achievement and success. The goal is no marking period grade below a C.

2.1 THE NEED FOR ACCOUNTABILITY IN THE CLASSROOM TEACHING-LEARNING INTERFACE _____

From time to time, you read or hear about the need for accountability in the classroom. Upon closer scrutiny, it becomes apparent that accountability has many facets. There is teacher accountability for designing and presenting to students a mode of classroom instruction that motivates them and encourages achievement. There is student accountability for taking advantage of the learning environment presented and putting forth the work and effort necessary to reap the positive rewards of achievement. There is parent/guardian accountability for seeing that the student puts forth the required effort for achievement both in and out of the classroom. And, of course, there is administrative accountability to provide for a classroom that is environmentally healthy and supplied with the proper teaching materials as well as adequate teacher preparation time.

In terms of accountability, there is much that you can do to design and initiate your own program. Further, it does not require a great expenditure of time and energy to so do. You need only a philosophical commitment that learning should be designed to encourage student achievement, and be designed in a fail-safe way that almost precludes student failure (assuming that the student is accountable by taking an active, serious interest in his or her success). Once you have made this commitment, you must make students and parents or guardians fully aware of your accountability commitment as well as what their own commitments entail. Once you accomplish these tasks, teacher–student rapport in the classroom can reach a new level of positive commitment.

2.2 THE RESULT OF ACCOUNTABILITY IN THE CLASSROOM TEACHING-LEARNING INTERFACE _____

The accountability approach to educating students results in a very positive, close three-way rapport between teachers, students, and parents or guardians. The main burden of learning, in the final analysis, falls squarely on the student. Gone are the multitude of a student's excuses for failure which tend to implicate everyone but the student him or herself. If and when a student fails, accountability makes it clear that the student is the one at fault. Thus, if at the end of the school year a student is going to fail a course, it comes as no surprise to the student or parent/guardian. Further, the student will recognize that the failure has occurred not by chance, but by his or her personal choice.

2.3   INITIATING A SUCCESS-DIRECTED LEARNING (SDL) PROGRAM OF ACCOUNTABILITY IN THE CLASSROOM ____

Success-directed learning (SDL) has several unique components:

a. The student has the chance to be retested on material.

b. The student is able to choose between oral and written tests.

c. A comprehensive after-school clinic is available to students who seek help in developing good study habits.

d. The student has a personal course achievement and grade average log.

e. The student learns in detail about SDL and accountability responsibilities.

f. Each student has a student cumulative folder containing the student's course work accomplishments.

g. Parents and guardians are introduced to SDL.

2.4   THE STUDENT'S INTRODUCTION TO SUCCESS-DIRECTED LEARNING ____

Use Worksheets 2.4–1 through 2.4–6 as a ready-made packet of handouts (or as guides in customizing your own version of SDL education), and distribute them to students. Go over this material carefully and in detail with students.

Instruct students to take the packets home for their parents or guardians to read. Remind them that they are to return to you, as soon as possible, the form (Worksheet 2.4–1) signed by a parent or guardian.

If you decide to try a success-directed learning program of accountability in your own classroom, you will be amply rewarded for your efforts as your students begin to demonstrate real motivation for learning.

Name: _____   Date: _____

A Parent's Introduction to SDL

Dear Parent or Guardian:

I currently have your son and/or daughter in my science class.

In teaching the class, I am using an approach to education that I call Success-Directed Learning, or SDL for short. *Success directed* means that the teaching techniques used have been designed to enhance the student's success in the course.

Success-directed learning is an accountability approach to education that emphasizes a close three-way rapport between teacher, student, and parent/guardian. The accountability roles for teacher, student, and parent/guardian are clearly and specifically spelled out in this packet of materials.

For your role in accountability, I would appreciate your doing the following:

a. reading this packet of information, which includes:
   • SDL (Success-Directed Learning): A Student's Introduction
   • An SDL Guide for Studying
   • SDL Testing
   • The SDL Thursday After-School Study Techniques Sessions
   • SDL Course Achievement and Grade Average Log.
b. making certain that your son/daughter follows the task of home studying as spelled out in this packet of materials.
c. reviewing the Course Achievement and Grade Average Log on a regular basis.
d. calling me at _____ at any time to discuss student progress.
e. filling out the form at the bottom of this letter and returning it to me via your son/daughter.

Thank you very much for your help.

Sincerely,

— —

I have read the materials that my son/daughter brought home to me.

_____

name of parent/guardian

phone # where I can be reached: _____

WORKSHEET 2.4–2

SDL (Success-Directed Learning): A Student's Introduction

1. What is Success-Directed Learning?

Success-directed learning is an accountability approach to teaching and learning that involves a commitment on the part of everyone that has an interest in your success. Your teacher has an interest in your success. Your parents/guardians have an interest in your success. And you yourself have an interest in your success. The commitment on everyone's part simply is to work together to assure your success.

2. Specific Techniques of Study

You will learn some very specific techniques of studying that will decrease your study time while increasing the amount of material learned and improving your memory of the material. As a result, you will earn higher grades on tests. (Please refer to the sheet, "An SDL Guide for Studying.")

3. The Taking of Tests

Tests may be taken in either the written or oral modes. If you feel that you might be better able to achieve on oral rather than written tests, they can easily be arranged. (Please refer to the sheet, "The Taking and Retaking of Tests.")

4. Retesting

You may be retested on material once if you are not satisfied with the original grade earned. Please refer to the sheet, "SDL (Success Directed Learning) Testing."

5. The Course Achievement and Grade Average Log

So that you can be aware of your grade average at all times, you are responsible for keeping this log up to date. As a result, your end-of-marking-period grade will never "come as a surprise." Further, the log should be reviewed from time to time by your parents/guardians. Log sheets will be distributed at the beginning of each marking period. Extra log sheets are always available.

6. **After-School Study Techniques Sessions**

Thursdays after school have been set aside specifically for improving study techniques. These sessions are designed to help the following two groups of students:

Group 1: Students who are earning below a C average. It is extremely important that these students take advantage of the study techniques sessions.

Group 2: Any student who desires a higher grade average and/or more efficient study techniques.

WORKSHEET 2.4–3

An SDL (Success-Directed Learning) Guide for Studying

An average study time of 30 minutes an evening should adequately prepare you for learning the information presented in this course. However, the use of proper study techniques is mandatory for

a. achieving a good understanding of the materials and concepts presented
b. earning good grades (a C or above) on all tests.

If you are experiencing difficulty achieving the above goals, please be aware of the Thursday after-school study techniques sessions. These sessions can be of considerable value to you, and can result in dramatic improvement in the understanding of the materials studied and the exams. These sessions are your avenue to earning the highest possible grades that you desire.

What follows is a detailed step-by-step study technique procedure that can be of significant value to you.

The major task of this study technique is to put all material to be learned on 3 × 5 cards. This includes:

a. class lecture notes
b. film, video, and film strip notes
c. study guide notes
d. and so on.

In preparing your 3 × 5 study cards, write a question on one side and the answer on the other side. It is important that you limit each card to one question and its answer.

The following example is that of preparing some 3 × 5 cards from material given as class notes. Assume that you have taken the following class notes:

> The human ear has three main divisions. They are the outer ear, the middle ear, and the inner ear. The outer ear is composed of two structures, namely the auricle and the auditory canal. The function of the auricle is that of directing sound waves into the auditory canal. The task of the auditory canal is to channel the sound waves to the ear drum. The ear drum separates the outer ear from the middle ear. The formal name for the ear drum is the tympanic membrane.

Using the 3 × 5 study card format, you could convert your notes to the following questions and answers. (See Worksheet 2.4–4.)

QUESTION ON FRONT OF CARD	ANSWER ON BACK OF CARD
What are the three main divisions of the human ear?	outer middle inner
What are the two structures of the outer ear?	auricle auditory canal
What is the function of the auricle?	to direct sound waves into the auditory canal
What is the function of the auditory canal?	to channel sounds to the ear drum
The ear drum separates what two divisions of the ear?	the outer ear from the middle ear
What is the formal name for the ear drum?	the tympanic membrane

Obviously, using your 3 × 5 study cards effectively involves using them properly. For example, at times you might wish to number your 3 × 5 cards in order, when the material is such that learning in a logical order will help facilitate memory storage in your brain.

One very important reason why using 3 × 5 study cards is far better than studying pages of notes is that the use of the cards keeps the mind active at all times. How many times have you been reading over and over again pages of notes only to realize that your mind has been wandering? This does not happen when you use the study cards. After reading each question, your mind is active in attempting to come up with the answer.

The use of 3 × 5 study cards has many additional positive factors that make it an ideal study technique:

1. Learning has been reinforced through the simple task of actually writing down the questions and the answers.
2. You get positive feedback every time you find that you have answered a question accurately.
3. 3 × 5 study cards can be easily carried with you, allowing for quick reviews.
4. You can flip through the study cards many times in a short period of time, allowing for many repetitions of exposure to the material. This enhances the brain's long-term memory.

For maximum benefits when using the study cards, the following procedures should be used to help the brain process the material:

1. Spaced study has been shown to be much better than massed study. What this means is that after going through your study cards two or three times, you should spend a period of time doing something else (perhaps turning to another subject), after which you return to the cards. Following this spaced study format will speed up the learning process.
2. As you use the study cards, read the questions and answers out loud.
3. Whenever possible, use visual imagery. For example, suppose the question is: What is the function of the auricle of the ear? When giving the answer, which is to direct sound waves into the auditory canal, picture the auricle actually directing the sound waves into the auditory canal.

Finally, your daily 30-minute study routine can be as follows:

1. Review previously prepared study cards to make certain that you fully know the material.
2. Prepare new study cards.
3. Learn your newly prepared study cards.

To help you study efficiently, you should consider setting up an area in your home that you can call your home study center. This area might even be a small section of your bedroom. Wherever the study area is located, it should be used for study purposes only. This adds a measure of seriousness to your studying and is of great value psychologically.

It is important that you set aside time from your busy daily schedule for studying. Granted, this can be difficult if you are trying, for example, to hold down a part-time job. However, you must make a commitment to schedule this study time. On a separate piece of paper, take time to organize a schedule that allows for this study time.

WORKSHEET 2.4–4

SDL (Success-Directed Learning) Testing

1. Any test taken during the year may be retaken (a *new set of questions* on the same material) once in an attempt to raise your grade, showing that you have increased your mastery of the course subject material.

2. A retest must be accomplished within one week of the in-class test. Please schedule the retake with your teacher.

3. A retest can be taken in either of the following two modes:
   a. Written
   b. Oral

4. There is one requirement for taking a retest: You must have all of the required subject matter prepared using the proper 3 × 5 study card format. These cards must be brought along with you at the time of the retest.

5. After retesting, the grade you receive will be an average of the initial test grade and the retake test grade. The following are two examples:

   ONE: a. initial test grade = F
                   b. retake test grade = A
                   c. grade received = C

   TWO: a. initial test grade = C
                   b. retake test grade = A
                   c. grade received = B

   NOTE: Obviously it is a much wiser use of your time to study efficiently for the initial test. Certainly a retest can significantly raise your grade, but you will have spent much more additional time in preparing for it, time that probably could have ensured an A on the initial test.

6. If, after taking a retest, your grade remains below a C, it is highly advisable that you begin attending the Thursday after-school study techniques sessions specifically designed to help assure that your grade average will be a C or better.

Name: _____   **Date:** _____

SDL Course Achievement and Grade Average Log

It is your responsibility to keep an accurate and updated log on your progress throughout the year in this course. (Your teacher has extra sheets if you need them.)

It is also your responsibility to show this log to your parents/guardians (preferably after every test grade is received) to keep them updated on your progress in this course.

Marking Period #: _____

TEST #: _____

GRADE ACHIEVED: _____

RETAKE GRADE (if applicable): _____

GRADE(S) RECEIVED ON ANY EXTRA CREDIT*: _____

Grade average achieved thus far: _____ (If grade average is below a C, refer to your sheet on Thursday study techniques sessions.)

TEST #: _____

GRADE ACHIEVED: _____

RETAKE GRADE (if applicable): _____

GRADE(S) RECEIVED ON ANY EXTRA CREDIT*: _____

Grade average achieved thus far: _____ (If grade average is below a C, refer to your sheet on Thursday study techniques sessions.)

TEST #: _____

GRADE ACHIEVED: _____

RETAKE GRADE (if applicable): _____

GRADE(S) RECEIVED ON ANY EXTRA CREDIT*: _____

Grade average achieved thus far: _____ (If grade average is below a C, refer to your sheet on Thursday study techniques sessions.)

---

*   Extra credit becomes effective only when your test grade average is a C or better. If your test grade average is below a C, the most important thing to work on is study techniques. You can, however, accumulate extra credit, which will "kick in" once your test grade average reaches a C or better.

Name: _____   Date: _____

The SDL Thursday After-School Study Techniques Sessions

The basic philosophy which led to the establishment of the Thursday after-school study techniques sessions as part of the success-directed learning approach is as follows:

> It is a rare student who actually desires to fail at his or her attempt at achieving one of the most important goals in life, that of a good education. When a student is having difficulty achieving in a course, it is rare that the student lacks the essential ability, but rather because he or she lacks the proper techniques of study.

1. The Thursday after-school study techniques sessions occur on a weekly basis.

2. A student is *expected* to attend the study techniques sessions if the following situation prevails:

    The grade average for the course is below a C.

    The student is *expected* to continue attending the study techniques sessions until:

    The grade average for the course is at least a C.

3. A student with a grade average of a C or above is *encouraged* to attend the study techniques sessions if:

    A higher grade is desired

    OR

    more efficient study techniques are desired.

4. At the study techniques sessions, specific techniques of study are presented in an easily mastered format. These study techniques, when employed in earnest, will significantly raise one's grade average through increased efficiency in learning the course material.

5. *Note:* A student who is achieving below a C average in the course and is not attending the Thursday study techniques sessions will have his or her guidance counselor notified so that a meeting can be arranged with parents/guardians to discuss the situation.

SECTION 3

General Classroom Management Techniques

Section 3 offers solutions to a variety of classroom management issues, ranging from dealing with a student who does not want to dissect and teaching evolution, to handling student tardiness and seating arrangements. Included are 15 related worksheets to help you implement the suggested techniques.

3.1 DEALING WITH THE STUDENT WHO DOES NOT WANT TO DISSECT _____

Today more than ever, people are concerned about the humane treatment of animals, whether in zoos, pet stores, or research laboratories. Similarly, many people question the necessity of dissecting preserved animals in the classroom. Is the educational value of dissection worth sacrificing animals' lives? Could not the teacher perform demonstration dissections, thus eliminating the need for classroom quantities of preserved specimens? Could videotapes be purchased (or made by the teacher) of dissections of selected organisms?

When a student tells you that he or she does not want to participate in a classroom dissection activity, you should discuss the situation with the student in private. Do not deal with the student in the context of the classroom setting by saying something like, "Dissection is an important part of this course and every student must participate in it." This approach is far too dogmatic, will probably embarrass the student, and will set the stage for an emotional confrontation between you and the student. Moreover, if the parents or guardians agree with the student, the entire school system's dissection policy may be called into question.

When you discuss the situation privately with the student, one of the following reasons for not wanting to dissect will probably emerge:

- The student does not wish to perform the dissection, but will observe and take the required notes. There is no reason why the student cannot participate on his or her own terms. There is nothing sacrosanct about the student experiencing dissection firsthand. Today even some medical schools allow a student to participate in dissection as an observer.
- The student does not wish to participate actively or passively in a dissection activity because of a personal religious or ethical belief. In this case, contact the parents or guardians and discuss the situation with them to determine the validity of the student's rationale. In the final analysis, it is the decision of the parents or guardians that should prevail. Perhaps an alternative activity for the student can be agreed on. There is no lack of library research topics that the student could explore. The student could prepare a paper on the role of the animal (being dissected in the classroom) in nature. Or a paper could be written on the care of animals in zoos. Further, if the student will not participate in an identification test using actual animal parts, plan to test that student using pictures of animal parts or his or her research paper.
- By refusing to dissect, the student is trying to avoid work. If you feel that the real motive for the student not wanting to dissect is to avoid work, you should discuss this with the parents or guardians and perhaps the guidance counselor. If avoiding work is the true motive, then when you

assign an alternative library research paper, the student will probably reconsider.

Inevitably, students will want to know where the specimens come from and how they are obtained. Many have pets at home and are sincerely concerned out of a reverence for living things. It would be ideal if you could say with certainty that the specimens were not raised and sacrificed solely for dissection purposes. For example, if you obtain hearts, brains, and the like from local abattoirs, you can explain to students that these specimens are indeed byproducts which otherwise would be discarded. As for cats and similar animals, unfortunately the answer is not as clear cut.

If you have developed a course handbook for your students, you should state clearly therein the extent of dissection that will be carried out during the year. If you have no such handbook, at the beginning of the course you should hand out a written statement concerning all dissections to be undertaken. In addition, if your school provides a course selection booklet for students, be certain that your course description includes the details of any dissections to be carried out.

3.2   TEACHING EVOLUTION AND/OR CREATIONISM _____

Most teachers know that when they teach a unit about the origin of life on Earth, they must be careful about how they approach the subject. The theory of evolution is not the only paradigm that deals with how life came to be on this planet. An alternate choice is referred to as the theory of creationism. In addition, there are those who find an eclectic approach more to their intellectual liking, one that combines the theories of evolution and creationism. These people believe in the evolutionary process, but believe that it is an example of a creator's handiwork.

It is important that you tell students that there are two main theories about life's origins, evolutionism and creationism. Tell them that as theories, each is subject to the scrutiny of scientific investigation (specifically to examine the factual evidence that has given rise to each theory). Emphasize to students that a theory develops out of accumulated facts. If there are no facts to back up a theory, one does not have a true theory, but has only a beginning hypothesis.

There will be students who hold one particular belief and absolutely refuse to listen to any evidence supporting an alternative belief. This attitude is, of course, antithetical to the nature and spirit of science. Tell students that any theory held as a personal belief system is only as strong as the factual evidence used in supporting it.

One excellent way to supplement a unit on the origin of life is to invite speakers of opposing viewpoints into your classroom. Select competent individuals who are knowledgeable relative to their own particular beliefs. Consider having two opposing speakers in the classroom at the same time to discuss their theories in a forum setting. Worksheet 3.2–1 contains a letter format that could be used in inviting guest speakers to the classroom. The letter is most effectively used following a personal phone call to the potential guest speaker. Worksheets 3.2–2 and 3.2–3 are designed to help students organize information (presented during class lecture and/or speaker presentation) relevant to evolutionism and creationism.

There may be parents or guardians who do not wish their son or daughter exposed to any theory other than what is taught at home. Rather than fight this situation (which can rapidly develop into a battle), it is probably better to provide the student with alternative assignments such as preparing a paper dealing with his or her own belief system regarding the origin of life. Further, you can give this student the opportunity to present his or her information to the class. This approach usually works well. In any event, Worksheet 3.2–1 contains a letter format that you can use in notifying parents or guardians about the unit on the origin of life.

There are two aspects of evolution that students are often confused about, whether or not they believe in the theory. The first aspect deals with the claim that humans directly descended from apes. The theory actually claims that humans and apes both evolved from a common ancestor. The second aspect deals with the concept that all life on Earth came from a common ancestor. For example, students often cannot grasp the concept that today's animals, which differ so greatly in appearance, could have all evolved from a common, animal-like life form. The development of the human body is an excellent analogy to use in helping students to understand the concept of evolution in action. Draw pictures on the chalkboard (or show actual pictures) of several types of cells, such as nerve, muscle, and blood, that make up the human body. Point out how totally different the cells look from each other. Then point out that all of the cells making up the human body developed (evolved) from a single cell (the zygote) that looks nothing like any of them.

Letter Formats for Inviting Guests and Notifying Parents or Guardians

I.   Inviting Guests

Dear _____:

    Thank you for being willing to come into my science class to discuss your views on the _____ theory of the origin of life. Also present will be _____, who will be discussing his/her views on the _____ theory of the origin of life.

    In teaching the origin of life, I am endeavoring to present fairly to the students the main points of each theory as well as any documented evidence that exists to back up each point. I sincerely appreciate your participating in this classroom program.

    The date of the program is _____, the classroom number is _____, and the time is from _____ to _____.

    Sincerely,

II. Notifying Parents or Guardians

Dear _____:

    In science class we are about to begin a unit on the origin of life. As you are probably aware, there are two main theories used to account for the origin of life forms. These theories are creationism and evolutionism. It is my sincere endeavor to present a fair picture of each of these theories. To this end, I have arranged for two individuals to come into the classroom as guest speakers to present their viewpoints on each of the theories.

    The individuals are _____ speaking on the theory of _____ and _____ speaking on the theory of _____.

    If you have any questions whatsoever, please do not hesitate to call me at _____ (phone number).

    Sincerely,

WORKSHEET 3.2–2

Information Data Table for Notes on Creationism

Instructions: As important points are presented, list them in the left-hand column and fill in the information asked for in the right-hand column.

Important points presented to explain why creationism is the correct theory to explain the origin of life.	Specific pieces of evidence that support each point.
1.	
2.	
3.	
4.	
5.	
6.	
7.	

Name: _____   Date: _____

Information Data Table for
Notes on Evolutionism

Instructions: As important points are presented, list them in the left-hand column and fill in the information asked for in the right-hand column.

Important points presented to explain why evolutionism is the correct theory to explain the origin of life.	Specific pieces of evidence that support each point.
1.	
2.	
3.	
4.	
5.	
6.	
7.	

3.3 SAFETY IN THE LABORATORY _____

We are living in an extremely litigious society. Thus you must be constantly aware of potential safety hazards during even the simplest of laboratory activities.

At the beginning of each school year, you should present a detailed unit on safety in the laboratory setting. Materials for building a unit on safety can be found in virtually all biological science supply house catalogs, such as the following:

> Ward's Natural Science Establishment, Inc.
> 5100 West Henrietta Road
> P.O. Box 92912
> Rochester, NY 14692-9012
>
> Carolina Biological Supply Company
> 2700 York Road
> Burlington, NC 27215

Instructional materials that can be purchased include 35-mm transparency sets, filmstrip/cassette programs, video programs, laboratory safety charts/signs/posters, guides to hazardous materials and their safe storage, and much more.

In addition, you should customize a safety unit to account for safety concerns unique to your particular classroom and/or laboratory program.

The following are some ideas that you could use in creating a unit on laboratory safety:

1. Have students make safety posters. These posters can be made on oaktag using eye-catching colors. They might contain messages such as: Never Forget to Wear Protective Eye Goggles; Do You Have Your Apron On?; Wash Your Hands Thoroughly after Labs; Throw All Wastes in Proper Containers; Do You Know Where the Fire Blanket Is?

2. Make certain that students know where to locate and how to use such items as the fire blanket, fire extinguisher, and eye wash.

3. Have students memorize an emergency signal. The emergency signal could be something as simple as blinking the classroom lights on and off. Students should realize that if the emergency signal is given, they should immediately stop what they are doing and pay close attention to you for further instructions.

4. Have students take a test on laboratory safety. After you have presented the instructional unit on safety, inform students that they must pass a test on the material prior to participating in any laboratory activities. Allow students to retake the test if they do not pass it the first time. Keep copies of the students' test results for future use if needed.

5. Make posters based on actual laboratory accidents. For example, assume that a student does not continually stir the hot water in a beaker after adding powdered agar and, as a result, the beaker breaks. A poster could be prepared with the actual broken beaker attached (the base of the beaker will show blackened agar where the agar settled, expanded, and shattered the beaker). The poster message should emphasize the importance of following laboratory instructions and wearing safety equipment such as goggles and aprons.

Worksheet 3.3–1 is a general checklist that you can use to determine the safety preparedness of a classroom laboratory. The list is in no way conclusive or definitive.

3.4   PREVENTING TEACHER BURNOUT _____

The beginning teacher and burnout:

Teacher burnout can occur at any time during your professional career. We tend to think of it happening to the teacher who has been teaching for a long time. However, it happens all too frequently to young, bright, talented individuals who, early in their teaching career, conclude that they are caught up in an educational system that does far more to stifle teacher ingenuity and creativity than to encourage it. In too many cases this perception is accurate. For these people burnout is rapid and complete.

Beginning to teach in any school system can be difficult for a new teacher. Some school systems have adapted a type of peer-guidance program for beginning teachers. These programs assign to the beginning teacher a more experienced teacher (the latter, hopefully, is an exemplary instructor). The beginning teacher needs someone to work with daily, one on one, especially concerning classroom discipline. The new teacher must realize that there is always a period of adjustment between teacher and students. Furthermore, the role of disciplinarian is entirely new to the beginning teacher. However, there is nothing as damaging as poor classroom discipline at the onset. It can lead to a quick burnout.

Burnout case scenarios and their attributes:

There are at least three basic scenarios that a teacher (or anyone in any occupation) can encounter relative to burnout. The first, the ideal-case scenario, involves the teacher who has never had a sincere desire to flee the classroom. This teacher has never had burnout and will tell you that every day (or almost every day) in a classroom filled with students is a challenge worthy of undertaking.

The second, somewhat less than ideal scenario involves the teacher who has always sincerely enjoyed teaching, but feels that it might be time for a change of occupations.

Classroom Laboratory Safety Preparedness

QUESTION	YES/NO

1. Do you teach a unit in laboratory safety? _____

2. Do you have a special signal to get the attention of the class in the event of an emergency? _____

3. Do you try out all laboratory demonstrations prior to class? _____

4. When presenting a potentially dangerous demonstration,

    a. is a safety shield used? _____

    b. do the students wear goggles and aprons? _____

5. Are ALL chemicals, biologicals, and acids kept in a storage area that assures no student access? _____

6. Are biological specimens kept in plastic shatterproof specimen jars? _____

7. Does the classroom have special ventilation to remove noxious odors during dissections? _____

8. Are safety cans used for dispensing flammable and combustible liquids? _____

9. Is the room equipped with an eyewash station? _____

10. Is the room equipped with a drench shower? _____

11. Is the room equipped with a fire extinguisher? _____

12. Is the room equipped with a fire blanket? _____

13. Are acid and alkali neutralizing kits available for emergency clean-up of acid or alkali spills? _____

14. Are autoclavable bio-hazard disposable bags used in getting rid of biological wastes such as bacteria? _____

15. Are there specially marked containers in the classroom for disposing of various types of wastes such as broken glass, liquids, and solids? _____

16. Are bars of soap and paper towels available for student clean-up after lab activities? _____

17. Are there any potentially hazardous areas in the classroom in terms of student movement and/or congestion during lab activities? _____

This teacher will tell you that the fun seems to have gone out of teaching; classroom instruction, for example, has become too routine and a bit boring, yet the course syllabus does not allow for change.

The third, the worst-case scenario, affects the well-being of both the teacher and the students. This teacher will tell you seriously that "personal meltdown" seems inevitable. He or she will exhibit symptoms of teacher burnout, including being "absolutely and truly fed up" with many of the following situations (situations, interestingly, that the non-burnout teachers also complain about, but successfully adapt to): (1) student assemblies being announced to teachers on the same day that they are to occur (or a day in advance); (2) homeroom announcements, over the school public address (PA) system, to which students initially listen until they realize (usually by the end of the second or third homeroom) that the announcements will rarely contain any information of substance; (3) PA system announcements "from the front office" that interrupt the day's instructional periods; (4) interruptions during class instruction in which passes are delivered to students requesting them to report immediately to the guidance office, or the nurse, or an administrator; (5) students leaving prior to the end of class instruction for sporting events and field trips; and (6) student discipline getting more difficult to manage. The list could go on and on.

Ideas for preventing and dealing with teacher burnout:

There is no panacea for teacher burnout. However, if you talk to teachers who have taught without burnout for 25 or 30 years, you may find out about teaching approaches and/or styles that tend to ward off major burnout crises. The following are some ideas:

1. *A teacher goal mental set:* Keep in mind that the primary goal of the teacher is to teach students. Never lose track of this goal. It can be repeated like a mantra when environmental conditions and events seem to dictate otherwise.

2. *On ending the workday:* There must be a time during the day when your workday ends and your home life begins. There must be a point at which you say, "What does not get finished today will have to wait until tomorrow." Otherwise, there will never be an end to the things that you feel you must accomplish *today*. There are teachers who feel guilty or somewhat less than dedicated if they go home empty handed. However, being overzealous can eventually evolve into feeling overburdened. Even if they leave work at school at the end of the workday, teachers will still accomplish on time, will have had a break from it, and will deal with it with refreshed vigor the next day.

There is an excellent mental technique that can help you leave behind all problems of the day when heading home. It works like this. When leaving school and driving out through gates to a main thoroughfare, the gates can represent the division between your workday and your home life. When you pass through the gates, you mentally shut out all of your workday problems and frustrations and open your mind to the enjoyment of home life ahead. By repeating this process on a daily basis (many teachers literally repeat it aloud to themselves), you will actually feel workday pressures and tensions lift as you pass through the gate. (Obviously, if a gate does not exist, any landmark can be used.)

Separating your workday from your private life is a critical factor in preventing burnout.

3. *The teacher–course relationship:* The classroom is your area of expertise in motivating students, establishing a pleasant rapport with students, and presenting students with learning opportunities using a wide variety of techniques. However, you may often feel that the cards are stacked against you in many ways. Perhaps, for example, you are never certain from one year to the next whether you will be teaching the same subject. This can happen when departments (usually influenced by an administration that feels it is not desirable to develop "teacher ownership of a course") feel that there must be teacher rotation of courses. The creative teacher is a very productive teacher in terms of constantly updating information taught, and redesigning and customizing materials for his or her unique teaching style. Obviously, this takes a lot of work and time. Of what long-term advantage is this creative output if it must be repeated the following year for an entirely new course? It is like constantly running to stay in the same place, and it can lead to teacher burnout.

Unfortunately, the aforementioned situation cannot be resolved easily unless it becomes department philosophy that a particular teacher has a right to choose to teach a particular course. For example, a teacher who develops and begins teaching a course in Field Ecology would have the option of continuing to teach that course if he or she so desires. This means that all of the teacher's effort and creativity in developing the course syllabus will pay off year after year. Now the teacher is no longer running to stay in the same place, but is running constantly to improve his or her course.

Once a link is established between a teacher and a course, many positive things happen. The teacher takes pride in his or her course. Efforts spent on developing the course in terms of content and teaching technique are not in vain. The teacher experiences an increased feeling of self-worth and pride because he or she has been entrusted with a particular science course. Teacher-course relationships can be of significant value in preventing burnout.

4. *An image of the teacher's role in the classroom:* It is important that a teacher have a positive self-image in terms of his or her role in the classroom. It is an important role, a tough role, a fun role, and a rewarding role. Teacher self-image might be described as follows:

A career in teaching can be considered akin to a career in the legitimate theater. Each class for the teacher is the staging of a performance for a student audience. The teacher is, to paraphrase a Hollywood term, a poly-hyphenate. He or she is the producer-scriptwriter-editor-special effects artist-director-performer-critic of every lesson plan.

As a producer, the teacher is responsible for supervising the entire classroom production (including classroom discipline).

As a scriptwriter, the teacher writes the scripts for all lesson plans with a format in mind that he or she thinks will capture and hold audience interest while at the same time developing in a logical manner the concept being taught.

As special effects artist, the teacher might use various visual effects such as computer graphics, overhead and opaque projections of charts and graphs, live demonstrations with living animals, and so on.

The teacher is a director in the true sense of the term, by having to sequence the interplay of lecture, student discussion, and audiovisual material. He or she must also encourage student rehearsals (home study) of script materials (from both text and lecture).

Performance is a crucial aspect in the art of teaching. The teacher as performer must motivate and stimulate classroom learning. A great performance is the icing on the cake of the previous scriptwriting and special effects efforts.

After the performance, it is time for the critics' reviews to come in. There are several sources of these reviews, and all are important for future course development. One source is student-audience reaction and feedback, perhaps based on such factors as attention span, questions asked, and participation in discussion. Another source is feedback provided by student quiz or examination grades. A third source is the teacher, who must decide, based on feedback from the other two sources, if editing and script rewrite are now necessary.

When a teacher incorporates into his or her self-image the roles actually played in the classroom to turn out a single, high-quality lesson plan, pride in accomplishment will follow. Combine pride in accomplishment with the concept of teacher–course relationship, and you have a very effective arsenal of arms against teacher burnout.

3.5   HANDLING THE PARENT/GUARDIAN CONFERENCE _____

Following are a few simple procedures that can be carried out to assure you a professionally confident and virtually stress-free approach to the parent/guardian conference. In most cases, if these procedures are followed, you will have very few, if any, conferences whereby the parent or guardian has to come into school, most being accomplished over the phone.

Parents and guardians are, almost without exception, extremely receptive to any action on your part that is directed to the welfare of their child. They appreciate notes sent home and/or telephone calls received (not necessarily just for informing them of problems; the purpose could be to note a grade improvement or other positive action) because it puts them directly into contact with their child's teacher.

Further, by doing so you are acknowledging them as an integral factor in their child's progress.

You can adhere to the following steps in initiating and moving forward toward the parent/guardian conference:

1. Whatever the reason for the conference, be it for a classroom behavior problem or an academic problem, discuss it personally with the student first. Explain that if the problem is not taken care of immediately, then a conference with the parent or guardian will become a necessity. Most students do not want their parents or guardians involved in a conference, and this is the only motivation they need to begin solving the problem.

2. In the event that a parent or guardian conference seems imminent, it is advisable first to send a conference request letter (Worksheet 3.5–1) to the parent or guardian. (Keep a copy of the form for your own records as documentation of this correspondence.)

3. Immediately follow up the mailing of the conference request letter with a phone call to the student's home. This puts you in personal contact with the parent or guardian, and underlines your sincere concern about the student. If the problem is one of academics, be certain to explain your study clinic minicourse (discussed in 1.12). This explanation by itself could eliminate the need for the parent or guardian to come into school for a conference. If the problem is one of behavior, the parent or guardian will almost always back you up on a remedy—

Teacher-to-Parent Letters of Communication

I. A Conference Request Letter:

Use of this letter will show your serious intentions in solving the problem. Further, it documents the effort you have put forth should the problem worsen as the year progresses. Therefore, keep copies for your records.

The following is a sample letter that has been used in communicating with parents and guardians.

Dear Mr. and Mrs. Granger:

I am sending you this note to inform you that Jerry is experiencing some academic problems in General Biology.

Although Jerry appears to be paying attention during class, I sometimes feel that his mind is miles away from the topic at hand. His test scores are beginning to drop. I have spoken to him about my concerns, but I do not think it is having an effect.

I think it advisable at this time that we consider planning for a conference during which we can sit down and discuss any remedial action that we feel should be taken regarding Jerry's circumstances.

I will be telephoning you so that we can discuss the matter further.

If you wish, I can be contacted at telephone number:

Sincerely,

II. A Letter of Commendation:

The following letter is included to encourage communication with parents and guardians regarding good news about their child. This aspect of communication is often overlooked, and yet it is important and not too time consuming.

Dear Mr. and Mrs. Marshall:

I am happy to be able to write you this note reporting good news about Wendy. Over this past marking period her grade average in General Biology has gone from a C– to a solid A. Her homework assignments and test results provide ample evidence that she is effectively studying the material.

I have congratulated Wendy on her rather dramatic improvement. She says that the animal units we have studied are really interesting to her and that she thinks she would like to work with animals in a future career.

Sincerely,

4. for example, a suggested seat change, or detention assignment. (Then, when making the seat change or assigning the detention, tell the student that this is a combined action of you and the student's parent or guardian.)

5. If it becomes necessary for one or more of the student's parents or guardians to come into school for a conference, it is desirable to have a third party, most often the student's guidance counselor, sit in with you during the meeting. If the problem regards academics, present the cumulative student folder (discussed in 1.17). Follow up with an explanation of your study clinic minicourse. These two actions literally place the obligation to initiate problem solving in the hands of the parent or guardian and the student. (It cannot be overemphasized that the combination of the cumulative student folder, used in documenting a student academic problem, and the study clinic minicourse, used in correcting the problem, are invaluable and powerful tools in communicating with parents and working with students.) If the problem is one of behavior, consider using the student contract approach (discussed in 3.6).

3.6   HANDLING THE CHRONIC DISCIPLINE PROBLEM: THE STUDENT CONTRACT

There are students who exhibit less than desirable classroom behavior. In some instances their behavior pattern seems to be chronic and does not respond adequately to simple admonishments or to the suggestion of a parent/guardian conference. Sometimes these students seem almost incorrigible. When dealing with students of this nature, the student contract, while not a panacea, can play an important role in helping to ameliorate the problem.

A sample of the student contract is shown in Worksheet 3.6–1, and it is used in the following context:

1. A conference is held with the student in attendance, along with his or her guidance counselor and parent or guardian.
2. The student contract is discussed so that it is understood by all.
3. The contents of the student requirement section and the default section of the contract must be decided on and written in during the conference.
4. Listed under the student requirement section should be those specific behaviors that are expected from the student while in the classroom. The list will depend, of course, on the problem, and could include such things as no talking privileges and not leaving one's seat.
5. Listed under the default section should be those specific things that will result if the student does not follow the dictates of the contract. They

could include anything from being removed from the class and receiving an F, to repeating the course, perhaps during summer school.

6. The student contract is signed by the teacher, guidance counselor, parent or guardian, and the student prior to ending the conference.

The Student Contract

The following contract will list specific basic behaviors expected of the student while in the classroom. It also lists the specific actions that will result if the contract is not followed.

Initially, the student will have the progress report section of this contract signed on a daily basis by the teacher and parent or guardian. If after seven school days the contract has been followed successfully, it need only be signed on a weekly basis for three weeks. If after three weeks the contract is still being followed successfully, it can be terminated by the student.

THE STUDENT REQUIREMENT SECTION:
    The following behaviors are required on the part of: _____
                                                                                (student name)
1.
2.
3.
4.

THE STUDENT DEFAULT SECTION:
    The following will result if the above behaviors are not exhibited:
1.
2.
3.
4.

                         SIGNATURES   Date: _____
Student:
Parent or Guardian:
Counselor:
Teacher:

THE PROGRESS REPORT SECTION:
School Day #   Teacher Signature and Comment   Parent/Guardian Signature
  1.
  2.
  3.
  4.
  5.
  6.
  7.

Week #
  1.
  2.
  3.

3.7 HANDLING STUDENT LATENESS TO CLASS

Experienced teachers know that students will be as lax in their classroom behavior as you are lax in your discipline policy. The same is true concerning students being on time for class. Some teachers unfortunately adopt one of two extremes. The first is allowing students continually to come into class up to four and five minutes late without a pass. Needless to say, students entering late disrupt the class.

At the other extreme is the teacher who announces that all students must always, without exception, be in class before the late bell rings, and if not they must always have passes or else detentions will be assigned. The problem with this approach is that there are, from time to time, valid reasons why a student might not be on time and might not have a pass. If a detention is assigned, the student may resent the teacher as well as the detention.

A moderate approach can eliminate the problems associated with policy extremes regarding student lateness to class. You can explain to students that when the classroom door is closed, any student entering will need a pass. You can judge when to close the door (perhaps 20 or 30 seconds after the late bell). The important point is that having this bit of built-in flexibility regarding student lateness will reduce resentment and hassles.

A beneficial technique is to make a little sign for the outside of your classroom door reminding students of the lateness policy. The sign might read as follows:

> THE CLASSROOM DOOR IS CLOSED AND YOU
> ARE NOW OFFICIALLY LATE TO CLASS.
>
> UNLESS YOU HAVE A PASS, A DETENTION
> WILL BE ASSIGNED.

Far from being intimidated by the sign, students will know that policy is being adhered to. They also will realize that they did have a brief grace period before the door was closed.

3.8 HANDLING STUDENT REQUESTS TO GO TO THE NURSE, RESTROOM, TELEPHONE, AND SO ON

It can be very frustrating when, during lesson instruction, a student requests permission to leave the classroom. Yet to make a blanket statement to the class that under no circumstances may a student leave the classroom is unrealistic. Therefore, you should have a policy for handling these requests, and you should make certain that students fully understand it. In classes where students are seriously interested in learning, there will virtually never be an abuse of the pass policy. However, when students are not serious about learning, they will invariably take advantage of any pass system, if for no reason other than disrupting and/or getting out of class. There are a few simple techniques that you can use to discourage requests. Consider the following policy guidelines:

- Encourage students to take care of such chores as using the restroom and making an important telephone call between class periods. The student who thinks he or she might be late for class as a result of such a chore should obtain a pass from you beforehand. Otherwise, if late without a pass, a detention will be assigned.
- Inform students that during the class period, a request pass will only be signed for an emergency visit to the nurse. If the student considers a restroom pass to be an emergency, let that student use the restroom at the nurse's office. The fact that the student will have to use the nurse's office will all but eliminate student restroom requests.
- Each time a student requests an emergency pass, have that student sign a paper with his or her name, date, and reason for the emergency pass. This allows you to keep a record of such emergency passes, and to quickly identify the student who might be playing games to get out of class. Further, the fact that the student is aware that you are keeping a record of such passes does much to discourage future requests. In addition, with the drug problem that exists in many schools, it is more important than ever to monitor student requests for leaving the classroom.
- Once you have documented data that a particular student is prone to repeatedly requesting emergency passes (for example, for a headache), then the parents or guardians, or the student's guidance counselor or an administrator might be notified to look more closely into the problem.

If you need to tighten up on your classroom pass policy, the preceding guidelines can be of great value.

3.9   KEEPING CURRENT IN SUBJECT MATTER CONTENT _____

Scheduling sufficient time to deal with day-to-day teaching tasks is an accomplishment in itself, not to mention keeping current in subject matter content. Your daily work as a teacher, rather than diminishing over the years, actually increases. Part of the reason for this is that during the first year or two of teaching, you are not familiar with the texts and lab manuals, and are mainly concerned with just staying a step or two ahead of the students. In addition, as a beginning teacher you spend time preparing instructional materials on a level suitable for students, because it is quite a transition from learning materials on the college level to teaching materials on the high school level. As the years pass, however, you will have more time to devote to enhancing and expanding the basic curriculum with supplemental materials. This latter task should be never ending, because it is the pathway to keeping instructional materials current with research and literature.

The following are ways of keeping current in subject matter content:

1. If you happen to be in close proximity to a college or university campus, take advantage of the many public seminars and lectures that are presented on various science topics.

2. If you have time during the summer, many colleges and universities present programs sponsored by various science and environmental organizations and designed specifically for the classroom biology teacher. Often, significant stipends are available. The biology department should keep you up-to-date on these seminar offerings.

3. Having speakers from various areas of biological science come into the classroom exposes both you and the students to current scientific information.

4. If you are near a college or university bookstore, up-to-date texts (and often less expensive secondhand ones) can be purchased to study.

5. You can take notes on television science programs viewed at home for later use to supplement classroom instruction.

6. Within the bounds of videotaping and copyright law, taped science programs and/or classroom quantities of science articles from magazines can be used to supplement classroom instruction. Both, of course, should be accompanied by student study guides.

7. Magazines are an excellent source of continuous up-to-date information on a wide variety of biology topics, and should definitely be taken advantage of. The following is a list of several selected magazines, along with addresses, that can be of value to biology teachers. Perhaps your school library subscribes to some of them.

NAME OF MAGAZINE	ADDRESS
The Sciences	The Sciences Subscription Department 2 East 63rd Street New York, NY 10131-0164
Longevity	Longevity P.O. Box 3226 Harlan, IA 51593-2406
The American Biology Teacher	The American Biology Teacher National Association of Biology Teachers 11250 Roger Bacon Drive #19 Reston, VA 22090

Scientific American	Scientific American Inc. 415 Madison Avenue New York, NY 10017
Science World	Scholastic Inc. 351 Garver Road P.O. Box 2700 Monroe, OH 45050-2700
Discover	Discover P.O. Box 420087 Palm Coast, FL 32142-9944
Science News	Science News P.O. Box 1925 Marion, OH 43305
National Geographic	National Geographic Society P.O. Box 2895 Washington, DC 20077-9960
Environment	Heldef Publications Environment 1319 Eighteenth St., NW Washington, DC 20077-6117
American Health	American Health P.O. Box 3016 Harlan, IA 51593-2107
Health	Health P.O. Box 52431 Boulder, CO 80321-2431

As teachers of science realize, the classroom text is usually the least up-to-date source of current happenings and investigations in biology. Use of alternative sources, particularly magazines, will go a long way in assuring that the information you impart to students will be current. It can be argued that the best classroom curriculum uses alternative sources for main content and the classroom text as a supplement.

3.10 PARENT'S NIGHT AND IDEAL TEACHER-PARENT RAPPORT

Many school systems have a yearly program in which parents and guardians are invited and encouraged to come into school to meet and talk with their children's teachers. These programs are called Parent's Night, Parent's Day, Back to School

Night, and so on. Whatever the name, this is an important event that can be of significant value to both teachers and parents or guardians. The following information can help you in preparing a Back to School Night program. It is based on a format that allows you approximately 10 minutes to make each presentation to each "class of parents" (we will include guardians under the term *parents*) as they follow a day in their son's or daughter's schedule.

1. Make sure your name and subject taught are written on the chalkboard before parents arrive.
2. Introduce yourself orally to parents in as pleasant a manner as possible. Remember, this is the first time most of the parents have ever seen you, and for better or worse, first impressions are being formed about you. Now and throughout your 10-minute presentation, your facial expression, tone of voice, and physical attire are all playing a role in establishing you as a competent professional.
3. Point out to parents the career posters (discussed in 1.8). Parents are usually interested in what career choices their sons or daughters have made.
4. If you keep student cumulative folders (discussed in 1.17), inform parents that they can at any time ask to look through their son's or daughter's folder. Parents will appreciate being able to do so.
5. Distribute copies of any course textbooks and lab manuals.
6. Distribute copies of your student course handbook (discussed in 1.13) for parents to take home with them.
7. Use an overhead projector for projecting transparencies of key pages of the student course handbook. Such key pages include an introduction to, or description of, the program; an overview of the contents covered in the course; the testing and grading policy; extra-credit work available; speakers and/or field trips; after-school study clinics; homework assignment expectations; and so on.
8. If you use VOEs (voluntary oral exams, discussed in 1.4) and/or The Steps to an A Game (discussed in 1.11), explain these motivational techniques to parents. They will be pleased and impressed.
9. Show a few slides of student edible models (discussed in 1.6) and explain how they are used as a motivation technique in the classroom. This is an excellent way to end your presentation on a high note.
10. Close your presentation by encouraging parents to contact you at any time during the school year if you can be of any help whatsoever.

Examples of classroom work can add a lot to Back to School Night. For example, you can set up a few microscopes along with a brief written explanation of what the microscope is focused on; and you can display student projects such as models or posters and classroom study models such as skeletons.

Even though there is only a 10-minute time allocation for making presentations, it is better to have a bit more material than you need rather than not enough material. It is surprising how much material can be presented in 10 minutes.

Remember, Back to School Night is the only time you will come into contact with a majority of the parents. You will not remember each parent that attended, but each parent that attended will remember you.

3.11   SEATING ARRANGEMENTS FOR STUDENTS _____

The Physical Arrangement of Chairs or Desks

The typical classroom has student chairs, or desks, arranged in rows. Twenty-eight chairs, for example, are probably arranged in four rows of seven each, or some such similar pattern. To add a unique touch to your classroom, all you have to do is design a different seating arrangement.

Allowing students to take an active part in choosing a seating arrangement is just another way to help create an ideal rapport between you and the students, especially at the beginning of the year. In addition, it is interesting and fun. You might begin by proposing various patterns, by diagramming them on the chalkboard or on an overhead projector transparency.

Another technique of choosing a seating arrangement is to allow students to design the patterns.

Regardless of which ploys you use in choosing a seating arrangement, the pros and cons of each arrangement (ease of seeing the chalkboard, and so on) should be discussed prior to the final decision.

Worksheet 3.11–1 diagrams four possible seating arrangements. In this case laboratory desks are used. A favorite among students, and possibly the most ideal of all the arrangements, is shown in section "a" of the worksheet. Rectangular in design, it allows all of the students to be in visual contact with each other. This arrangement is particularly advantageous for student discussions and debates. Further, it allows for a closer proximity to the students if you utilize the open rectangular space for moving about.

Another popular seating arrangement is shown in section "b" of the worksheet. This arrangement is great for courses during which there is a lot of lab work requiring small group teamwork.

Preparing the Class Seating Chart

You have two options in preparing the class seating chart: (1) assigned seating, and (2) random seating. If you choose assigned seating, it can be based on alphabetical order or on the strategic placement of students to minimize possible future discipline problems. In either event, the student has no input about where he or she sits.

On the other hand, you might elect to allow students to sit where they wish, by whom they wish. Of course, students like this option best, and it is another way to help create, right at the beginning of the year, an ideal rapport between you and the students. When using this option, however, you must emphasize to students that selecting their own seats is a privilege that can and will be revoked in the event of inattentiveness.

If a student uses poor judgment in seat selection and is a disruptive influence on others, that student's seat should be immediately changed. However, the change need not be irreversible. The student can be moved with the understanding that if the disruptive behavior ceases, he or she will eventually be moved back to the original seat choice. This process of rewarding the student for exhibiting desired classroom behavior is an excellent one and should be used often.

Worksheet 3.11–2 is an example of a seating chart. A convenient way of dealing with a seating chart is to prepare it on the cover of a manila folder. You can keep test papers, anecdotal notes, and other materials specifically for that class in the manila folder. If you are allowing students to select their own seats, it is advisable not to fill in the seating chart for the first few days. During this time the students will often try out various seats. Finally, when you do fill in the seating chart, use pencil. In this way, if a student's seat is changed, you can change it easily on the chart.

3.12   SELF-EVALUATION TECHNIQUES

During the course of the school year, you will undergo formal classroom evaluations by an administrator and/or department head. For the most part, the results of these evaluations become part of your permanent record, both for the district and the state. A major concern, however, is that the evaluations are limited not only in terms of how often they are carried out over the course of the school year, but also in the scope of what they evaluate. You are observed teaching only two or three lessons out of hundreds taught over the course of the school year. There are different teaching styles, ranging from the inquiry mode to the lecture mode, each requiring a different type of

Four Examples of Classroom Seating Arrangements

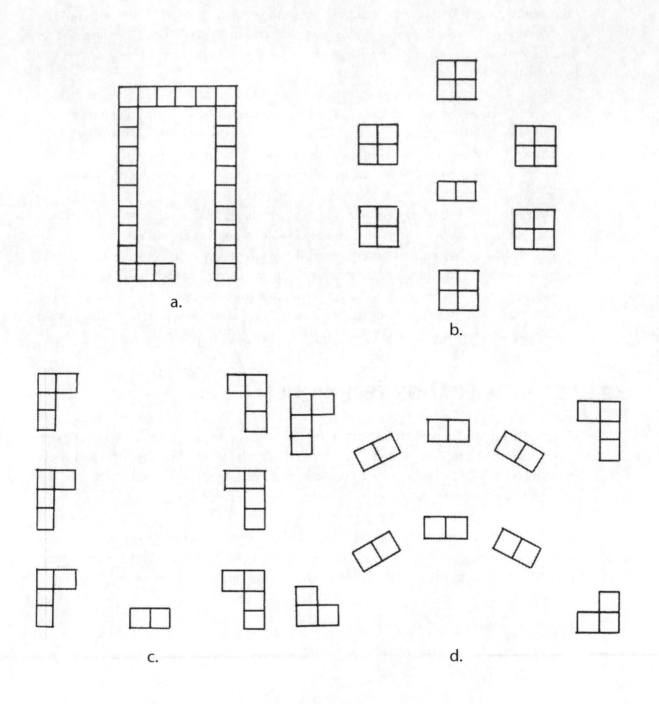

a.

b.

c.

d.

A Sample Classroom Seating Chart
(desks in room arranged in a rectangle)

CLASS: _____

\# IN CLASS: _____

PERIOD: _____

evaluation instrument. The reliability (accuracy and consistency over a period of time) and validity (evaluating what it is supposed to evaluate) of any evaluation instrument used to rate teachers should be verified.

Before any formal observation, it would be ideal if the following prerequisites could be met:

1. The teacher should have a copy of the evaluation instrument well in advance of the actual evaluation.
2. The evaluation instrument should have been tested for reliability and validity.
3. It should be clear to the teacher what categories of behaviors on his or her part are going to be observed.
4. It should be clear to the teacher what behaviors will be judged acceptable and not acceptable.
5. The evaluator should have had formal training in evaluation.
6. The teacher should be able to have a conference with the evaluator within two or three days of the evaluation.
7. The teacher should receive a copy of the prepared written evaluation.
8. The teacher should have the right to write a formal rebuttal in response to anything that he or she feels was unfair about the evaluation process.

If you are concerned about your classroom performance, you should consider self and/or peer evaluation as a valuable tool. You can personally design the evaluation instrument to reflect exactly what classroom behaviors concern you. Not only can this type of evaluation be pressure free and fun, but it forces you to focus on your classroom behavior.

Self-evaluation can be carried out with a television camera-recorder (camcorder). Or, if you have a particular teacher-friend with whom you share a close rapport, you might ask this peer to observe you and fill out the evaluation instrument. Each of these approaches to evaluation has its own advantages. A peer may fill out the evaluation instrument more objectively than you. On the other hand, observing yourself on videotape can be a real eye opener. A combination of both modes is advisable.

The steps in organizing and carrying out self or peer evaluation might be as follows:

1. Prepare an evaluation instrument that is designed to rate you according to the specific classroom behaviors, based on your own teaching mode, that you are trying to achieve. (This evaluation instrument can be used repeatedly. Worksheet 3.12–1 is an example of an actual teacher-made instrument. It contains those items that this particular teacher was concerned about regarding her general classroom teaching. The

evaluation instrument that any teacher prepares is custom made based on the unique concerns of that teacher.)

2. Prepare a hypothetical evaluation summary that you feel will be the actual outcome of the evaluation. (Worksheet 3.12–2 is an example of a hypothetical evaluation summary.)

3. If using a peer, meet with that person and go over your evaluation instrument, detailing exactly which behaviors you are most concerned with.

4. Have your peer come into your classroom and use the evaluation instrument.

An Example of An Evaluation Instrument

DATE OF EVALUATION:

CLASS:

LESSON:

Specific behaviors to be evaluated:

1. Is voice modulation (as opposed to monotone) in evidence?

2. Does the teacher articulate words clearly?

3. Does teacher present a neat appearance?

4. Can teacher's voice be heard throughout the room?

5. Does teacher move about the room during the lesson (as opposed to staying in one general location)?

6. Does teacher display interest in the subject matter being taught?

7. Does teacher appear to enjoy teaching the class?

8. Are the students paying attention to the tasks on hand?

9. Did teacher begin the class on time (as opposed to three or four minutes into the period)?

10. Did instruction continue until the end-of-period bell (as opposed to student "free time" during the final three or four minutes)?

11. Does teacher frequently monitor student learning by asking questions?

12. When a student responds correctly to a question, does teacher give positive reinforcement, through a smile or a "well done," etc.?

13. When a student gives a wrong answer does the teacher avoid giving evidence of negative feedback?

14. Does teacher frequently encourage the students to ask questions that they might have on their minds?

15. Does teacher pace the lesson presentation (as opposed to going too fast or too slow)?

16. If a discipline problem arises, does teacher handle it in a "low key" way while maintaining self composure?

17. Would you say that a great majority of the lesson was supplemented by visual aids?

18. Was the teacher's handwriting on chalkboard, etc. clearly legible?

Hypothetical Evaluation Summary
(teacher notes in boldface––selected behaviors only)

DATE OF EVALUATION:

CLASS:

LESSON:

1. Is voice modulation (as opposed to monotone) in evidence? **Probably not. Unless I keep it in mind, I tend to monotone.**

2. Does teacher articulate words clearly? **I think so.**

4. Can teacher's voice be heard throughout the classroom? **Many times when ending sentences or phrases, I tend to lower my voice a bit too much. Students often ask me to repeat.**

5. Does teacher move about the room during the lesson (as opposed to staying in one general location)? **I should do this more to make my presence felt, especially with this particular class.**

6. Does teacher appear to enjoy teaching the class? **I do enjoy it, but probably should show it more by smiling more. I tend to keep a neutral expression on my face.**

10. Did instruction continue until the end-of-period bell (as opposed to student "free time" during the final three or four minutes)? **I have to watch this. I tend to end the lessons a little too soon, especially as the year progresses.**

11. Does teacher frequently monitor student learning by asking questions? **I think the evaluation will show that I do this very well.**

13. When a student gives a wrong answer, does the teacher avoid giving evidence of negative feedback? **I do not think I give negative evidence, but I want to be sure.**

14. Does teacher frequently encourage the students to ask questions? **I do this a lot. However, I am probably guilty of not giving them enough time to formulate and ask the question. I tend to move on too fast.**

15. Does teacher pace the lesson presentation (as opposed to going too fast or too slow)? **I tend to go too fast. I have to slow down.**

17. Would you say that the great majority of the lesson was supplemented by visual aids? **Yes. I am strong in this area.**

18. Was teacher's handwriting on chalkboard, etc. clearly legible? **This is something I have to keep in mind constantly. I tend to write too fast. Often students ask what I have written.**

5. Compare the results that your peer recorded on the evaluation instrument with your hypothetical evaluation outcome.

6. If you prefer not to use a peer and you have access to a television camera and recorder, you can set up the equipment in your room to record your teaching of a lesson. You can arrange it in a stationary position, or you can have someone operate it for panning, and so on.

7. Using a videocassette player, evaluate your lesson using your evaluation instrument. Compare the results with your hypothetical evaluation outcome.

Another, more specific, type of evaluation instrument that can be designed is one that will help you if you wish to focus on a particular instructional mode, such as inquiry. Worksheet 3.12–3 is an example of a format that could be used in evaluating an inquiry approach to teaching. Worksheet 3.12–4 has been filled in with hypothetical data.

In an inquiry approach, you are not a dispenser of facts for students. Instead, students discover information for themselves. Your role is that of a catalyst in guiding students along pathways to discovery. You and students alike should pose a lot of questions to promote student discussion and understanding. Throughout the lesson, you will often probe (asking one or more follow-up questions after the original question) a student's understanding of the material.

Another helpful technique (but one that is seldom used because of teachers' time constraints) is to sit in on another teacher's class and watch that teacher teach. Some schools even give in-service credit for doing this. A workable situation is when two teachers agree to sit in on each other's classes. Classes need not be restricted to the field of science, but can and should be in other disciplines as well.

When sitting in on another teacher's class, it is probably best to refrain from taking actual notes, because this can be stressful to the teacher you are observing. Rely instead on mental notes. A lot of good techniques can be picked up from observing other teachers.

3.13 KEEPING INVENTORY OF MAJOR LABORATORY EQUIPMENT AND GENERAL SUPPLIES _____

All too often, inventory time (when the department chair announces that the laboratory equipment and general supplies list for next year's order is needed) seems to pounce on us unexpectedly. And all too often you wish that you had kept a running tab of equipment and general supplies needed. Of course, you resolve to keep this running tab beginning with the next school year. At the end of the school year many departments need yet another list, this time an updated list of laboratory equipment (along with serial or other identification numbers) for insurance coverage purposes. Worksheets 3.13–1, 3.13–2, and 3.13–3 will help make inventory more manageable.

Worksheets 3.13–1 and 3.13–2 allow you to keep an updated inventory of all major laboratory equipment and general supplies, respectively. Major laboratory equipment includes pieces of nonexpendable equipment that last for several years of use. General supplies, on the other hand, include such expendable items as chemicals (biologicals) and glassware, which often have to be replaced on a yearly basis.

An Evaluation Instrument for the Inquiry Instructional Mode

Instructions: Place an X in the appropriate column each time the following behaviors are displayed:

1. teacher asking question
2. teacher probing (teacher using a follow-up question to probe student understanding)
3. student responding to question (asked by teacher or another student)
4. student asking question (to teacher or another student)
5. teacher answering question (teacher directly answering a question rather than leading student to the answer)
6. teacher providing information (teacher providing factual material as in a lecture format)

teacher asking question	teacher probing	student responding to question	student asking question	teacher answering question	teacher providing information

Hypothetical Evaluation Data

1. Results of a lesson that is not true inquiry:

teacher asking question	teacher probing	student responding to question	student asking question	teacher answering question	teacher providing information
X X X	X X	X X	X X X X X X	X X X X X X	X X X X X X X X X X X X X X X

2. Results of a lesson that is true inquiry:

teacher asking question	teacher probing	student responding to question	student asking question	teacher answering question	teacher providing information
X X X X X X X X X X X X	X X X X X X X X X X X X	X X X X X X X X X X X X X X X X	X X X X X X X X X X X X X X X X	X X X	X X X

An Inventory of Major Laboratory Equipment

ROOM #:
TEACHER:
SCHOOL YEAR:

MAJOR LABORATORY EQUIPMENT					
ITEM NAME	SERIAL (ID)#	ITEM NAME	SERIAL (ID)#	ITEM NAME	SERIAL (ID)#

An Inventory of General Supplies

II—GLASSWARE

ITEM NAME	QUANTITY ON HAND	ITEM NAME	QUANTITY ON HAND	ITEM NAME	QUANTITY ON HAND

II—CHEMICALS (BIOLOGICALS)

ITEM NAME	QUANTITY ON HAND	ITEM NAME	QUANTITY ON HAND	ITEM NAME	QUANTITY ON HAND

Major Laboratory Equipment and General Supplies Needed for Next Year

1—MAJOR LABORATORY EQUIPMENT

ITEM NAME	QUANTITY NEEDED	COST PER UNIT	TOTAL COST	SUPPLIER	CATALOG #

Worksheet 3.13–3, when filled out on an ongoing basis during the school year, can save considerable time when the list of supplies needed must be turned in. It has been designed to record information needed for bid lists, which schools use to keep operating costs down.

3.14 PERSONAL EXPENDITURES RECORD TABLE _____

Although many biology departments provide a petty cash fund, it might cover expenses only up to, for example, $10. In addition, petty cash usually runs out well before the end of the school year. Therefore, many teachers, from time to time, purchase certain classroom curriculum-related materials using out-of-pocket money. Some of this out-of-pocket money is for classroom curriculum-related materials such as books and magazines (including subscriptions), expenses incurred in attending a seminar, and other items that are not necessarily covered by the school budget.

Worksheet 3.14 is a convenient table that can be used to keep track of out-of-pocket expenses for later reimbursement. Some of the data recorded in this table will prove beneficial at tax time, since many education-related materials that incur out-of-pocket expenses are tax deductible. For convenience, keep your personal expenditures record table in a manila folder along with all receipts.

3.15 THE ART AND SCIENCE OF TEACHING _____

Is there an art to teaching? If so, there must be certain skills that a prospective teacher should acquire through experiences and observations during teacher training. Is there a science to teaching? If so, there must be a specific body of knowledge that a prospective teacher should learn during teacher training to ensure that knowledge will be presented in a proper sequence conducive to learning.

The Art of Teaching

That prospective teachers must experience student teaching supports that teaching is partly an art. Much of the student teaching experience is concerned with observation of how an experienced teacher uses his or her skills in motivating and maintaining student interest levels during lessons. The presentation of the lesson itself (not the material presented) places the teacher in the role of a performer on a stage. A teacher who is expert in the art of performing uses deliberate body gesticulations and movements, facial expressions, and voice modulations in lesson presentations.

There is a right-brain approach to teaching students. Research suggests the right cerebral hemisphere of the cerebrum responds not to verbal stimuli, but to such non-verbal stimuli as movements and changes in sound intensities. These right-brain stimuli produce emotional responses in a person, in this case the student. (Think of a dynamic speaker that you have heard. This speaker probably did not speak in a monotone, standing motionless in one location. The speaker no doubt exhibited excellent voice modulation, hand and arm gesticulations, and body movement. Rock concerts rely heavily on right-brain stimulation. The performers are in constant movement about the stage. The music varies in loudness. Laser lights and pyrotechnics are liberally employed.) This right-brain approach to teaching is the art of teaching.

Personal Expenditures Record Table

DATE OF PURCHASE	ITEM	COST	REIMBURSABLE FROM PETTY CASH?	NOT REIMBURSABLE (FOR TAX RECORD)

The Science of Teaching

That teacher education contains courses dealing with such topics as The Methodology of Teaching and The Psychology of Learning supports that teaching is partly a science. For example, Bloom's Taxonomy (*Taxonomy of Educational Goal: The Classification of Educational Goals,* edited by Benjamin S. Bloom, New York: Longman, 1984) holds that there is a hierarchy of six increasingly complex levels of handling knowledge:

	LEVEL		BRIEF DESCRIPTION
1.	Knowledge	1.	Recalling previously learned information
2.	Comprehension	2.	Understanding information learned
3.	Application	3.	Applying learned information to new situations
4.	Analysis	4.	Breaking down learned material into component parts and understanding the relationship between the parts
5.	Synthesis	5.	Discovering new and creative relationships between different sets of learned material
6.	Evaluation	6.	Making judgments and drawing conclusions

There are curricula that have been designed in accordance with this hierarchy.

Whenever you praise a student for a specific behavior such as doing well on a test, you are using a technique based on scientific information resulting from investigations by psychologist B. F. Skinner. Skinner called this technique positive reinforcement. It simply means that a behavior that is immediately followed by a reward (verbal or otherwise) is reinforced.

Entire K–12 learning programs have been developed based on the work of the Swiss psychologist Jean Piaget. Piaget notes the following four stages of cognitive development:

	STAGE		BRIEF DESCRIPTION
1.	Sensorimotor	1.	Sucking, grasping, can find hidden objects (birth to age 2)
2.	Preoperational	2.	Symbolism, draw pictures, beginning of language (age 2 to 6)
3.	Concrete operational	3.	More complex intellectual activities such as simple math (age 6 to 12)
4.	Formal operational	4.	Abstract thinking, problem solving, value judgments (age 12 to adult)

The science aspect of teaching is not emphasized or discussed in many biology departments because of a lack of basic and applied educational research as well as a lack of widespread dissemination of research findings to teachers. Moreover, there are many high schools that do not have a professional library containing up-to-date sources of current educational research. Further, teachers do not have much time to keep up with research findings or to incorporate such findings into the curriculum.

Overall, a teacher's personal instructional skills call upon a combination of methodology derived from both art and science, and represent an interaction of skills used in several areas, including (1) knowledge and use of instructional materials, (2) knowledge of course content material, (3) planning, (4) management in the classroom, (5) knowledge of human growth and development, and (6) human relations.

SECTION 4

An Inquiry Approach to Experiencing the Spirit and Nature of Science

Section 4 presents an effective approach to teaching scientific inquiry (the spirit) and research technique (the nature), including 13 reproducible student worksheets. Be sure to read the list of important safety caution concerns for the Science As Science (S-A-S) units presented in this section before using them. These are detailed in 4.2.

4.1 A LIST OF THE MAJOR LABORATORY MATERIALS REQUIRED FOR TEACHING THE S-A-S UNITS _____

Ascorbic acid solution
Balloons
Beakers (assorted sizes)
Bean seeds
Bunsen burners
Cheesecloth
Cotton
Flasks (assorted sizes)
Fruit juices (assorted)
Fruits (assorted, fresh)
Glass marking pencils
Graduated cylinders (assorted sizes up to 1000 ml)
Indophenol (2,6-dichloroindophenol) 0.1%
Lab aprons
Lab balance
Lab scoops
Medicine droppers
Molasses
Pans (for water displacement)
Paper cups (12 oz., wax coated)
Paper towels
Plant food (2 brands of, water soluble)
Potting soil
Rulers (metric)
Safety goggles
Stirring rods
Stoppers (1 hole, compatible with glass tubing)
Test tubes (assorted sizes, including smaller ones that can be inverted into larger ones)
Tubing (glass with inner diameter of at least ¼ inch)
Tubing (rubber, compatible with the glass tubing)
Vitamin C tablets
Yeast (active dry, 1/4-oz. packages)

4.2 IMPORTANT SAFETY CAUTION CONCERNS FOR THE S-A-S UNITS _____

1. Students should be instructed how to insert glass tubing safely through one- hole stoppers. (One technique is to coat the glass tubing with

petroleum jelly and gently twist [never push] the glass tube through the hole of the stopper.)

2. Students should be instructed how to bend glass tubing safely. (One technique is to hold the ends of the piece of glass tubing so that the center of the tubing is over a Bunsen burner flame. Slowly turn the tubing so that the center is heated on all surfaces. As the center of the tubing begins to get red hot, it can easily be bent to the desired angle. The process of bending the tube should be gentle, not forced. The tubing should always be bent in a direction away from you and all others, in case the tubing breaks and shatters. Safety goggles and a lab apron should be worn by the student bending glass and by others in the vicinity.)

3. Students should be instructed how to ignite a Bunsen burner safely. (One technique is to always light the match and hold it over the mouth of the barrel of the Bunsen burner prior to turning on the gas.)

4. Students should be instructed how to put stoppers safely into flasks and test tubes. (One technique is to twist gently [never push] the stopper into the mouth of the flask or test tube until it is secure.)

5. During laboratory activities in which students are designing carbon dioxide collecting apparatuses, several will probably design closed systems. An example of a closed system is a flask, containing yeast solution and molasses-water mixture, with a glass tube passing out of it through a one-hole stopper through another one-hole stopper into another flask used to collect the gas. Closed systems are not safe because excess gas that might build up has no place to escape and thus may break the flasks. Open systems are much safer. An example of an open system is a water displacement system in which the gas is collected in an inverted graduated cylinder as the water is displaced into a pan. In this open system gas will bubble out into the water with no pressure build-up in the graduated cylinder. When a student presents a closed system design during a seminar, point out the dangers and use the opportunity to engage the class in a discussion regarding the necessity of designing safe experimental apparatuses. Before allowing any student to use his or her apparatus, make certain it is not a closed system.

6. During laboratory activities in which students are assembling carbon dioxide collecting apparatuses, caution students not to seal any part of the apparatus. For example, students might want to seal, perhaps with wax, around the edges of stoppers to help ensure a tight fit in test tubes, flasks, and so on. The reason for not sealing the system is to allow the stoppers to act as a safety device and pop out if excess gas pressure builds up inside the apparatus. Before allowing any student to use his or her apparatus, make certain that this safety caution has been followed.

7. Caution students that the tip of a gas collecting tube inserted into the container of yeast solution and molasses-water mixture should be

positioned well above the level of the liquid and never submerged into the liquid itself. The reason for this is that if the tip of the glass tube is submerged into the liquid, it cannot provide an exit for the gas produced and the gas may build up and break the container, a dangerous situation. Before allowing any student to use his or her apparatus, make certain that this safety caution has been followed.

8. Students must wear safety goggles and aprons at all times when working with their carbon dioxide collection apparatuses. There is always a remote chance that a piece of glassware might break due to gas pressure build-up.

9. As with any list of safety concerns, it is not possible to foresee every potential laboratory hazard. Therefore, you *must* read all of the laboratory activities carefully prior to using them in class. In this way you will become aware of any other safety concerns that might be important in your laboratory classroom.

4.3   AN INTRODUCTION––THE S-A-S APPROACH TO TEACHING SCIENCE

The S-A-S (Science As Science) Approach is a refreshing and very effective approach to teaching science that will bring fun and excitement to the teaching process. Teaching S-A-S means allowing your students to experience the spirit and nature of laboratory investigation mainly through using their cerebral matter and not cookbook instructional matter. In this section, the S-A-S Approach will be explained in detail and will be followed by several laboratory activities designed to familiarize you with this approach. Once familiar with S-A-S, you will be able to convert as much of your classroom program as you wish to this simple, efficient, yet seldom used approach.

An S-A-S Approach will provide the following:

- Much less lecture and much more lab
- Alternatives to lengthy fact-finding exams
- Techniques that allow you to keep close track of student understanding of classroom and/or laboratory activities
- An atmosphere of scientific inquiry in the lab
- Ample opportunity for independent student investigation
- Opportunity for you to become involved with students in the role of discovery
- Opportunity for students to master communication skills, both written and oral.

The spirit and nature of science

The spirit of science is inquiry, and the nature of science is experimental technique. When you teach a science course using the S-A-S approach, you are attempting to give students a realistic experience of what science is all about. You are trying to eliminate an approach based on lecture and rote memorization of facts, an approach devoid of thought and understanding.

The S-A-S Approach emphasizes reflective thinking on the part of students. It is a science-as-a-thought-process approach. It is steeped in student participation, which is approached from the standpoint of the student as a unique person working as an individual scientist solving problems while interacting with peers.

Through S-A-S the student will come to know science as the scientist knows it, and will see learning through problem solving as a personal endeavor that does not cease with the completion of formal education, but flourishes throughout his or her life.

The private and public aspects of science

There are two aspects of scientific endeavor: the private aspect and the public aspect. The private aspect includes the laboratory activities of the scientist. Whenever possible, the student should work individually, experiencing the scientific approach to problem solving including formulating hypotheses, designing experimental approaches, carrying out research, and collecting and analyzing data.

Public science involves rapport between the scientist and the lay public. This public aspect of science is brought into play when the student presents research findings to the rest of the class. It can be likened to a scientist explaining his or her research to the lay public. The student will present data orally, as in an oral presentation during a classroom seminar, or in written form, as in a science notebook.

The classroom seminar

The classroom seminar can become a microcourse in rhetoric dealing with all the skills necessary for clear, concise communication between human beings. In a seminar, students will have plenty of opportunities to develop their ability to speak and write clearly, concisely, and factually. The seminar is a clearinghouse for all that goes on in the laboratory. In addition to paper presentations, the seminar can deal with the discussion of such topics as laboratory techniques, procedural problems, and conclusions drawn from data.

The self-analysis query

The query is a question that the student answers in writing prior to giving an oral response. It can be used at any time: in the middle of a seminar; during a student's oral presentation; before, during, or after a lab. In addition to using queries for student self-analysis, you might want to use some for grading purposes. The query is used whenever you wish to get feedback from the class on a particular topic. Three examples follow:

> Query 1: Explain exactly how you used the 50-ml pipette in adding 10 ml of water to the yeast-molasses solution. (This query emphasizes the ability to express oneself clearly. The papers are exchanged and students attempt to follow the written directions while demonstrating with the pipette.)
>
> Query 2: Based only on the facts written on the chalkboard, write a conclusion. (The papers are collected and a few are projected on the classroom screen using an opaque projector. Class discussion should focus on the validity of the conclusions.)
>
> Query 3: You have just heard Debbie state that "in my molasses-yeast test tube, a lot of carbon dioxide was produced at 30 degrees." Comment on her statement. (Selected papers should be read as classroom discussion focuses on the need to be specific and to avoid ambiguities, such as "a lot" [how much?] and "30 degrees" [Fahrenheit or Centigrade?]. How much yeast in how much molasses?)

The query should be used often. It keeps students aware of current activity in the classroom and gives them the opportunity to practice expressing themselves in clear, concise language. It also allows you to be aware of who is and is not paying attention. You are going to be surprised and frustrated at times to find out that just when you are certain that everyone knows what is going on, query responses indicate otherwise.

4.4   A SAMPLE UNIT USING THE S-A-S APPROACH

The working format in using S-A-S is simple: A problem is posed for investigation. The rest is up to you and your students working together. Your role is that of a catalyst: You get the stage set and get things into action, guide students along when need be, and allow them to make mistakes as they find solutions. Students develop the hypotheses, outline experimental procedures, design the data tables, and draw the conclusions. It is the students' show. They are in the driver's seat of the scientist. You are along for the ride, keeping the car on track and directed toward the final solution of the problem.

You will be guided step by step through the entire sample S-A-S unit that follows. Worksheets are provided as guidelines. Some are designed to be handed out to students, while others are for your use in guiding students. You might wish to give students the choice of working individually or in teams of two.

SAFETY NOTE _____

Prior to beginning this unit, make certain that you have read and are familiar with the important safety cautions outlined at the beginning of this section in 4.2.

_____

A.   Students Receive a Copy of Worksheet S-A-S 1

This worksheet states the problem to be solved and provides space for note taking as you discuss with students the sequence of steps they should use in solving the problem. Make sure that students eventually have the following information:

STEPS	INFORMATION
1. Statement of the problem	This has already been stated above on the worksheet.
2. The procedure	A detailed, step-by-step explanation of how the investigation is going to be conducted. The ideal procedure is one that is written clearly enough that another scientist could conduct the investigation without asking questions.
3. The materials needed	A list of all of the materials (as well as quantities of each) needed to carry out the investigation
4. The data table	Designed and labeled so that all data eventually recorded in it is clear to any other scientist studying it
5. The analysis of data	A written interpretation of what the data means in terms of answering the problem being investigated
6. The conclusion	A summary statement based on the analysis of data that tells what you have learned in terms of an answer to the problem being investigated

Explain to students that their first task, as indicated on the worksheet, is to design an experimental procedure that they think will solve the molasses-yeast problem. Emphasize to them that the ideal procedure, in terms of being specific and clear, is one that another scientist could pick up, read, and carry out without having to ask questions.

WORKSHEET S-A-S 1

The Molasses-Yeast Problem

The following investigation is open ended, which means that you, as an inquiring scientist, will design your own unique experimental procedure to be followed in an attempt to solve the problem posed. After you have completed your procedure, you are going to present it orally to the class prior to carrying it out in the lab.

THE PROBLEM

You have been hired as a scientific consultant by a company that is in the business of producing and selling carbon dioxide gas. You are informed that it costs the company $1.00 for each 1 ml of yeast solution that is used. (The standard yeast solution that will be used is prepared by thoroughly mixing one 7-gram package of active dry yeast in 1000 ml of water at a temperature of approximately 110 degrees F.) It also costs the company $1.00 for each 1 ml of molasses that is used. The company wants you, the scientist, to determine the optimal combination of yeast solution and molasses-water mixture needed to produce the greatest amount of carbon dioxide gas (as measured in milliliters) and what the cost will be.

The Sequence of Steps That Should Be Used in Solving the Problem:

| STEPS | EXPLANATION |

1. _____ :

2. _____ :

3. _____ :

4. _____ :

5. _____ :

6. _____ :

B. Students Learn the Value of Keeping a Laboratory Log

Discuss with students the concept of the laboratory log. Use the following information as a guide to discussion:

As a scientist investigating and attempting a solution to a problem, you should keep a detailed log of everything that you do in this endeavor. It should be organized in a logical manner, and should be neat and clear enough so that another scientist could read it and know exactly what the state of your progress is.

ORGANIZATION OF THE LABORATORY LOG

1. Notes that should be recorded during each and every one of your laboratory investigations:
    a. The date of entry in the log
    b. The statement of the problem
    c. The procedure (plus a sketch of the experimental set-up)
    d. The materials used
    e. The data table
    f. The analysis of data
    g. The conclusion
    h. Additional notes that should be recorded when appropriate:
        - Were any problems encountered in carrying out the procedure?
        - Were any changes made in the procedure?
        - Were any problems encountered in taking data?
        - Did any ideas develop that might be used in the future?
2. Notes that should be recorded during each and every classroom seminar:
    a. The date of entry in the log
    b. The subject or purpose of the seminar
    c. All relevant notes that pertain to the subject being discussed. It is important that you give credit in your log to any individual or lab group that provides an original idea or technique that you incorporate in your own investigation. Giving credit where credit is due is an important part of the ethics of science.

C. Students Receive a Copy of Worksheets S-A-S 2 and 3

Worksheet S-A-S 2 is the format for the section of the student log that deals with laboratory investigations. Worksheet S-A-S 3 is the format for the classroom seminar section of the student log. You can either have students make additional copies of the formats on notepaper, or you can duplicate multiple copies of the worksheets for distribution. In any event, student use of the format will greatly facilitate your spot

Name: _____ Date: _____

WORKSHEET S-A-S 2

The Laboratory Log Format for
Laboratory Investigations

The statement of the problem:

1. The procedure (along with a sketch of the experimental set-up):

2. The materials used:

3. The data table(s):

4. The analysis of data:

5. The conclusion:

Additional notes that should be recorded when appropriate:
   a. Were any problems encountered in carrying out the procedure?
   b. Were any changes made in the procedure?
   c. Were any problems involved in taking data?
   d. Did any ideas develop that can be used in future investigations?

WORKSHEET S-A-S 3

The Laboratory Log Format for the Classroom Seminar

1. The subject or purpose of the seminar:

2. All relevant notes that pertain to the subject being discussed:

3. Any ideas or techniques that can be used in my own future investigations:

       Ideas or Techniques   Source of Credit

checking of logs. For example, after a classroom seminar you might wish to spot check item 2 in the seminar section of student logs. By spot checking only certain sections of logs, perhaps for grading purposes, you avoid hours spent reading each log in its entirety.

D. Students Experience Their First Classroom Seminar

Worksheet S-A-S 4 is not for student distribution. It outlines a few typical initial designs that students can be expected to come up with to investigate the molasses-yeast problem (discussed in Worksheet S-A-S 1), and you should use it as a guide for the seminar.

SAFETY NOTE _____

You will note that procedures 1 and 4 are closed systems. Closed systems were cautioned against in the important safety cautions listed at the beginning of this section. However, thin-walled balloons can stretch to accommodate gas as it is produced.

In this instance, the classroom seminar will be used for a discussion of the experimental procedures and equipment designs that various individuals or lab groups have come up with to solve the problem stated on Worksheet S-A-S 1. An excellent idea would be to have a student volunteer chair the seminar. (This reflects the spirit of the S-A-S Approach and is a good interaction experience for students. Of course, you are always there to guide and keep things on track. Remind students to record seminar notes in the appropriate section of the log.

The chairperson assumes a position at the front of the class and begins by reviewing with the class the purpose of the seminar. Then, perhaps by using the pic-a-tag technique (discussed in Section 1), the first presenter can be selected to present his or her experimental procedure.

Encourage the class to participate by reacting both pro and con to all experimental procedures and designs.

SAFETY NOTE _____

Refer to the list of important safety concerns presented at the beginning of this section.

This will give students plenty of practice in articulating and asking the "right kind" of question. Student participation keeps students interested and alert to what is happening in the classroom. Further, it gives the presenter(s) experience in responding to questions with clear, concise answers.

Ask a few questions of your own, strategically throughout the seminar. As the catalyst, you stimulate the proceedings if the going gets dull; you speed up the proceedings if they tend to slow down.

After four or five presentations have been made, a query can be used. In this instance the query could be: We are discussing experimental procedures to collect data to answer what question? Each student writes his or her answer on paper. Collect the papers before oral discussion of the query begins. During the discussion, selected papers can be projected on the classroom screen, allowing the class to comment on answers given.

An Example of Some Initial Experimental Procedures That Students Might Design for the Molasses-Yeast Problem

The procedures outlined below are purposefully crude to illustrate what will probably result from student first attempts at solving the problem. Students will probably omit such important factors as quantity of material used; size of glassware; how measurement of the quantity of gas produced is to be accomplished; the effects of elasticity of balloons; and so on. However, as research and seminars continue, students will become much more proficient at experimental design.

Procedure 1:

A. Fill a test tube with molasses.
B. Add 3 ml of the yeast solution.
C. Put a balloon over the mouth of the test tube to trap any gas produced.

Procedure 2:

A. Fill three test tubes with yeast solution.
B. Add different amounts of molasses and water mixture to each tube.
C. Invert small test tubes into the liquid in the larger test tubes to collect any gas produced.

Procedure 3:

A. Fill a beaker with a yeast and molasses mixture.
B. Invert a graduated cylinder (filled with yeast-molasses mixture) into the beaker to collect any gas by displacement.

Procedure 4:

A. Fill several test tubes with different amounts of molasses and water mixtures.
B. Add the same amount of the yeast solution to each test tube.
C. Collect gas by attaching a balloon over the mouth of each test tube.

Procedure 5:

A. Fill test tubes with molasses diluted with water.
B. Add yeast solution.
C. Using one-hole stoppers, glass tubing, and rubber tubing, channel any gas produced from the test tubes to inverted water displacement tubes.

At the conclusion of the seminar, inform students that their next step is to assemble their experimental apparatuses prior to carrying out their own individual investigations. Also, they should finalize their procedures in their logs, making any necessary changes or adding any ideas that they might have picked up during the seminar.

E. Students Request Materials Needed for Assembling the Apparatuses for Their First Molasses-Yeast Investigation

Distribute copies of Worksheet S-A-S 5. The forms on this worksheet will expedite distribution of laboratory materials needed for the upcoming investigation as well as those in the future. (Remember, several different lab teams are going to be approaching problems from different angles. Therefore, different lab teams will probably need different kinds of materials.)

Instruct students that for the upcoming investigation, as well as those that follow, they should use the slips provided on Worksheet S-A-S 5.

F. Students Carry Out Their Initial Investigation of the Molasses-Yeast Problem

You will be amazed at the variety of laboratory apparatus that the various lab teams set up to produce and collect carbon dioxide.

Your task is to move from lab station to lab station, **constantly checking the safety of the experimental apparatuses that students have assembled.** You should also be questioning the students: What are you doing? Why are you doing it? What data do you hope to obtain? What problem are you trying to solve? Why do you think that your particular procedure will solve the problem? Are you having any particular problems at this time? Are you aware of any potential safety concerns that must be considered as you use your apparatus?

(*Note:* Some students may use pure molasses instead of a molasses-water mixture stated in the problem. Do not correct them. Their data results will indicate little or no carbon dioxide production, and by the end of the next seminar they will have determined their own error. Similarly, some students may use balloons to trap the carbon dioxide. Eventually they will face the dilemma of how to measure the carbon dioxide trapped in the balloon. They too will realize their problem during the next seminar. The point is, let students learn from their own mistakes.)

In addition, monitor students to be sure that they are recording relevant information in their logs, such as a sketch of their experimental set-up.

G. Students Experience Their Second Seminar as They Discuss the Results of Their Initial Investigation

This seminar uses the same format as the first seminar. The various lab groups present reviews of their experimental procedure along with actual data collected.

By the conclusion of this seminar, students will be aware that a solution to the molasses-yeast problem is going to be considerably more involved than initially expected.

Name: _____   Date: _____

WORKSHEET S-A-S 5

Slips for Requesting Lab Materials

Student Name(s):   Date Needed:

Materials Needed:

Student Name(s):   Date Needed:

Materials Needed:

Student Name(s):   Date Needed:

Materials Needed:

Student Name(s):   Date Needed:

Materials Needed:

Student Name(s):   Date Needed:

Materials Needed:

Student Name(s):   Date Needed:

Materials Needed:

The following three concerns should be among those considered during the second seminar:

> *First concern: What is the best way to collect the carbon dioxide gas?* Based on the results of the first investigation, a consensus should be reached regarding the best apparatus to use. Are using balloons the best way? Are inverted graduated cylinders using the water displacement method best? What about small inverted test tubes inside larger test tubes?
>
> *Second concern: Is everyone being very specific in recording exact amounts of materials and sizes of glassware used in investigations?*
>
> *Third concern: What are the variables involved in the molasses-yeast investigation?* Use the following question–answer format as your guide in stimulating student discussion regarding the variables:

Q. Will someone review for us the problem that we are trying to solve?

A. To determine the optimal combination of yeast solution, molasses, and water to produce the greatest amount of carbon dioxide gas and how much it will cost.

Q. How many variables are we working with to produce the carbon dioxide?

A. Four. They are the yeast solution, the molasses, the water, and the carbon dioxide.

Q. How does the yeast solution figure in as a variable?

A. One could vary the amount of yeast solution added to the molasses-water mixture.

Q. How do the molasses and water figure in as variables?

A. Their amounts can be varied in the molasses-water mixture.

Q. How does the carbon dioxide figure in as a variable?

A. The amount of carbon dioxide produced will vary according to the amounts of yeast solution, molasses, and water used.

Q. Now that the variables are understood, what do you think the next step in your investigation should be?

A. To attempt to work with the variables one by one.

Q. What is an example of working with a single variable?

A. Varying the amount of yeast solution added to a series of test tubes containing a standard molasses-water mixture. Or adding a standard amount of yeast solution to varying concentrations of molasses-water mixtures.

Add to the preceding questions and modify them to fit the needs of your students. Remember, there is a lot of freewheeling using the S-A-S Approach.

Next, have students work together in designing two investigations. One investigation will be designed to study the molasses-water mixture variable, and the other investigation will be designed to study the yeast solution variable. Inform students that all information should be entered in their logs, because eventually they will be carrying out both investigations in the lab.

Worksheet S-A-S 6 is not for student distribution. It is for your information and provides examples of two workable procedures that have been used in investigating the molasses-water mixture variable and the yeast variable.

H. Students Investigate the Molasses-Water Mixture Variable and Then Present Their Findings in a Seminar

During seminar presentations, it will become obvious that the greatest amount of carbon dioxide will have been produced somewhere between the mid and lower percentages of the molasses-water mixture. If it turns out that it is the 10% mixture that produces the greatest quantity of carbon dioxide, let the seminar decide whether to accept 10% or to investigate molasses-water mixtures of less than 10%.

The percent mixture decided upon is the one to use in investigating the effects of variable amounts of the yeast solution on carbon dioxide production.

I. Students Investigate the Yeast Solution Variable and Then Present Their Findings in a Seminar

Let's assume that during this seminar it is established that the 10% molasses-water mixture containing 6 ml of the yeast solution produced the most carbon dioxide. The total cost is determined by adding the total milliliters of yeast solution in the test tube to the total milliliters of pure molasses (used in the molasses-water mixture) in the test tube and then multiplying the sum by $1.00. Thus, an answer has been obtained to the original problem.

The S-A-S Approach need not end here. Extra-credit incentive could be offered to those who wish to pursue the problem further (for example, in an attempt to produce carbon dioxide at a cheaper cost, or compare different brands of molasses).

4.5   A SECOND S-A-S UNIT

This second S-A-S unit uses the same format as described earlier for the sample unit. This time the S-A-S Approach is used as the vehicle for investigating the effects of two different brands of plant food upon the growth of bean plants.

A. The Student Seminar Introduces the Problem and Students Begin Designing Experimental Procedures

Worksheet S-A-S 7 is not for student distribution. It gives you an experimental design for investigating the problem that you are going to present to students during this seminar. Use the worksheet as a guide as you work with students during the next few periods.

You should select the two types of water-soluble plant foods that will be used in the investigation.

An Example of Two Basic Procedures That Students Might Design to Investigate the Two Variables Associated with the Yeast-Molasses Problem

In the procedures that follow, exact quantities of materials, ratios of materials, and sizes of glassware used will vary. In each of the procedures, the technique used to collect the carbon dioxide will depend on the consensus of opinion (usually water displacement) arrived at in the seminar.

Procedure 1: Investigating the effects of variable concentrations of the molasses-water mixture upon the production of carbon dioxide

A. Prepare the yeast solution according to the directions.
B. Label test tubes 1 to 11.
C. Fill the test tubes 3/4 with the following molasses-water mixtures:

Test tube	1:	100%	molasses
	2:	90%	molasses
	3:	80%	molasses
	4:	70%	molasses
	5:	60%	molasses
	6:	50%	molasses
	7:	40%	molasses
	8:	30%	molasses
	9:	20%	molasses
	10:	10%	molasses
	11:	100%	water

D. To each test tube add 4 ml of the yeast solution.

Procedure 2: Investigating the effects of variable amounts of yeast solution upon the production of carbon dioxide

A. Prepare the yeast solution according to directions.
B. Prepare a molasses-water mixture using the percent that was determined in Procedure 1 to have produced the greatest amount of carbon dioxide.
C. Label test tubes 1 to 10.
D. Fill the test tubes 3/4 with the molasses-water mixture.
E. Put 1 ml of the yeast solution into test tube 1, 2 ml into test tube 2, 3 ml into test tube 3, following in like manner through test tube 10.

An Example of an Experimental Procedure to Study the Effects of Two Plant Foods Upon the Growth of Bean Plants

Materials:

water

ruler (graduated in centimeters)

paper towels

glass stirring rod

two flasks

glass marking pencil

30 wax-coated 12-oz. paper cups or similar containers

potting soil (enough to fill the paper cups)

two different brands of water-soluble plant food

30 bean seeds

beaker

graduated cylinder

lab scoop

cotton

lab balance

Procedure:

1. Soak the bean seeds for approximately 24 hours in a beaker of water.
2. Fill each cup with moistened potting soil.
3. In each cup, plant a bean seed about 1.5 cm deep.
4. Divide the 30 cups into three groups of 10. Label the cups in two of the groups with the name of the plant food with which they will be treated. The third group (control) is labeled "no plant food."
5. Place the cups in a well-lit area, out of direct sunlight.
6. Check the cups on a daily basis and water as necessary (guard against soggy soil, as this will cause the seeds to rot).
7. Prepare the water-soluble plant foods according to directions on the package. Store the solution in labeled flasks, plugged with cotton, for future use.
8. As each bean seedling in the plant food groups reaches a height of 2 cm, begin moistening the soil once a week with the proper plant food solution, substituting it for the water. (The control group, however, still gets water.) Determine the millimeters of plant food solution needed for moistening the soil, and use this as the standard amount for adding liquid to all three plant groups.
9. During the five-week (or longer) growth period, record daily measurements of growth in height for each plant.

At the end of the growth period, record the following data:

a.   The average height of the plants in each group
b.   The total weight of the plants in each group
c.   The total weight of the stem and leaf portion of the plants in each group
d.   The total weight of the root portion of the plants in each group
e.   The total number of leaves of the plants in each group.

Write on the chalkboard the following basic problem to be researched, and have students copy it in their logs:

You have been employed by a plant grower to investigate the effects of two different plant foods [*write in the names of the plant foods you selected*] on the growth of bean plants from the time they break through the soil until they are at least five weeks old.

(*Note:* The preceding problem states "at least five weeks old" for plant growth; you might wish to continue for a longer period of time.)

Have students work in private lab teams of two or three as they spend the remainder of the seminar time designing their own investigations. (*Note:* It is crucial that each group does its own work based on its own ideas and not ideas from other groups. What you want from students are several different variations on investigating the problem.) Moving from group to group, you can monitor progress. Refrain from answering questions about, or commenting on, experimental design at this time.

At the end of the class period, you might want to collect the logs and review the various experimental procedures before continuing the seminar during the next class.

During the next class seminar, the lab teams should begin their presentations of experimental designs. Important factors in experimental design that the lab teams might have overlooked and must be guided into discussing are as follows:

- The number of seeds that should be used (Many will choose three, one for each plant food being tested. In this case, what happens if one plant dies?)
- Soaking seeds for 24 hours prior to planting
- The standard depth at which the seeds should be planted
- Schedules for watering the seeds
- A control group of plants
- When to start, and schedules for, adding the plant food
- How much plant food to add
- What data is going to be taken and when
- How weekends will be handled.

At the end of the class period, remind the groups to turn in request slips for any needed lab materials.

B. The Lab Teams Carry Out Their Laboratory Investigations of the Plant Foods

Make certain that for this lab, as well as all labs, students wear lab aprons and thoroughly wash their hands at the conclusion of the lab activity.

You might wish to soak the seeds prior to the lab period. Students can then proceed to plant them, label the containers, and so on.

If you do not have a well-lit location in which to keep the seedlings, you might have to use an artificial lighting arrangement.

For the first few days, the students' lab task will be to water the potting soil if it tends to dry out.

As the seedlings break through the soil and begin to grow in height, students will have to prepare the plant food solutions and begin their daily height measurements.

C. The Lab Teams Design Data Tables for Their Plant Food Investigations

Worksheet S-A-S 8 is not for student distribution. It gives you background to use as a guide in student discussion.

Have each lab team design its own data table and then present it to the class for comment and discussion.

The goal of the seminar is for students to understand the necessity of designing a data table that is logically organized and clearly labeled.

D. The Lab Teams Collect Their Final Plant Food Investigation Data

Following their collection of data on the final day of the investigation, direct students as they dispose of wastes in the proper containers and clean any equipment used.

E. The Lab Teams Compare and Discuss Their Data Findings

Have the lab teams organize their data collected and fill in their data tables. Then have them compare and discuss their data findings. Data findings from the various groups should be similar and in general agreement. If data conflicts or is unclear, a class discussion should follow in an attempt to explain and rectify the discrepancy.

4.6   A THIRD S-A-S UNIT

This third S-A-S unit suggests another laboratory activity using molasses and yeast, and deals with precision of laboratory technique and measurement. A simple statistic called the variance will be used in analyzing data.

SAFETY NOTE _____

Before beginning this laboratory activity, make certain that you read and become familiar with the list of important safety concerns outlined for the first molasses-yeast unit, "A Sample Unit Using the S-A-S Approach."

_____

A. The Student Seminar Introduces the Problem and the Class as a Whole Designs an Experimental Procedure

Write on the chalkboard the following problem to be researched, and have students copy it in their logs:

How much carbon dioxide in milliliters is produced over a 24-hour period using a 10% molasses solution with 5 ml of a standard yeast mixture added?

An Example of a Data Table for Recording Data Collected from a Study of the Effects of Two Plant Foods on the Growth of Bean Plants

Student data sheets will differ from the one that follows; however, you can use this one as a guide as you work with students.

I. Daily Measurement (cm) of Plant Growth in Height

Plant Food Group:	Plant Food Group:	No Plant Food
DAY HEIGHT	DAY HEIGHT	DAY HEIGHT
1 _____	1 _____	1 _____
2 _____	2 _____	2 _____
3 _____	3 _____	3 _____
ETC.	ETC.	ETC.

_____cm = the average height of plants treated with plant food:
_____cm = the average height of plants treated with plant food:
_____cm = the average height of plants not treated with plant food.

II. Measurement	Plant Food Group:	Plant Food Group:	No Plant Food
• Total # of Leaves on Plants			
• Total Weight (grams) of Plants			
• Total Weight (grams) of Stem and Leaf Portion of Plants			
• Total Weight (grams) of Root Portion of Plants			

Allow the class as a whole to design the experimental procedure to be used by each lab team in solving the problem. If you had students do the first S-A-S unit using molasses and yeast, their experience in that unit will help them now.

You should not intervene at all, but let students design the entire procedure even if you are aware that it has flaws. The major concern of this investigation is actually accuracy in mixing solutions and setting up laboratory materials in an attempt to achieve similar data results from the various lab teams.

Worksheet S-A-S 9 is not for student distribution. It outlines a crude type of experimental procedure that will probably be similar to the first one your students designed. It also outlines an experimental procedure that is specific in detail and similar to the one your students will develop as this unit progresses.

B. Students Carry Out the Investigation

Provide a package of active dry yeast and container of molasses for each lab team. It is essential in this investigation that each team prepares its own yeast-water solution and 10% molasses solution.

SAFETY NOTE _____

Before any student runs an investigation, check the safety of each experimental apparatus, making certain that there are no potentially hazardous apparatuses.

After 24 hours, the students take their data.

C. Students Compare Their Data Results in a Seminar

Have the seminar chairperson write the following column headings side by side on the chalkboard:

LAB TEAM #	ML OF CARBON DIOXIDE PRODUCED OVER A 24-HOUR PERIOD

The chairperson then fills in the columns as information is provided. It will be obvious that there are large variations in the amounts of carbon dioxide produced as reported by the different lab teams. This should lead to the main question to be posed in the seminar: Since each lab team is using the identical procedure in carrying out the investigation, why are there such large variations in the amounts of carbon dioxide produced? To solve this dilemma, encourage students to come up with as many factors as possible that might vary from lab team to lab team. Further, students should address ways in which these factors can be dealt with. The following factors might be among those discussed:

FACTOR	HOW TO DEAL WITH
• Different amounts of yeast in the different packages	• Prepare a standard yeast-water solution for the entire class.
• After the yeast is mixed with the water, it tends to settle to the bottom of the container.	• Thoroughly shake the container immediately before each use.
• When preparing the 10% molasses solution, accurate measurements are not obtained.	• Use 100-ml graduated cylinders for measuring amounts over 10 ml, and 10-ml graduated cylinders for measuring amounts under 10 ml.
• Not knowing the proper procedure for preparing a percent solution	• Use a ratio of 10 ml of molasses to 90 ml of water.
• Failure to read the graduated cylinder properly	• Read the meniscus when using the graduated cylinder.

Finally, all of the factors that students discuss should be incorporated into a new experimental procedure that can then be carried out.

D. The Seminar Continues as Students Are Introduced to the Statistical Concept of the Variance

Distribute copies of Worksheet S-A-S 10 and let students complete it. The worksheet is a self-explanatory introduction to a statistical concept known as the variance. In Part A, students are led through the step-by-step procedure using a hypothetical set of data. In Part B, students are instructed to calculate the variance for each of their two investigations to determine whether they have achieved a lower variation in the second one.

The calculations for Part A of the worksheet follow:

Lab Team #	Milliliters of Carbon Dioxide Produced over a 24-Hour Period	The Difference between the Average and Each Individual Measurement	The Differences Squared
1	250	107	11,449
2	60	83	6,889
3	100	43	1,849

4	175	32	1,024
5	40	103	10,609
6	340	197	38,809
7	70	73	5,329
8	110	33	1,089
9	225	82	6,724
10	60	83	6,889

The average = 143.

The sum of the fourth column = 90,660.

The calculation of the variance using the formula:

$$\text{The Variance} = \frac{90,660}{9}$$

The Variance = 10,073.

Note: The variance is a large number because the individual data measurements varied considerably. Your own class results from the first investigation could be similar.

Examples of Experimental Procedures Investigating Carbon Dioxide Production

The following rather crude experimental procedure is similar to that developed by students for their initial investigation of the problem.

1. Mix the yeast and water.

2. Prepare the 10% molasses solution.

3. Add the yeast solution to the molasses solution.

4. Collect the gas for 24 hours.

The following experimental procedure is similar to that developed by students as they learn from experience through lab investigations and seminar discussions. They will realize that the ideal experimental procedure leaves nothing to the imagination. Each step of the procedure should be so clear and specific that another scientist could follow the instructions without asking any questions.

1. Using a 1000-ml graduated cylinder, dissolve the contents of a package of active dry yeast in water according to the directions on the yeast package. (Pay close attention to the temperature requirements noted on the package.) Thoroughly mix the yeast with the water by cupping a hand over the mouth of the cylinder and carefully agitating the cylinder for three minutes.

2. Prepare a 10% molasses solution using a ratio of 10 ml of molasses to 90 ml of water. Stir the mixture thoroughly.

3. Pour 25 ml of the molasses solution into a test tube.

4. Using a 10-ml graduated cylinder, add 5 ml of the yeast solution to the molasses solution.

5. Using a one-hole rubber stopper, glass tubing, rubber tubing, and a 500-ml graduated cylinder, arrange the apparatus to collect the carbon dioxide by water displacement.

*SAFETY NOTE*_____

Be certain to show students how to put a piece of glass tubing safely through a one-hole rubber stopper.

_____

6. After 24 hours, record the milliliters of water that have been displaced (representing the milliliters of carbon dioxide produced) from the graduated cylinder.

WORKSHEET S-A-S 10

The Calculation of the Variance

Introduction:

You have already determined that your carbon dioxide data results vary quite a bit between the different lab teams. You have just finished designing a second investigation in an attempt to reduce this data variance. What is needed at this point is a way to measure how much the data varies between lab teams. Once you measure the variance in your first investigation, then you can determine whether or not you have indeed reduced the variance as a result of your second investigation. Variance can be defined as the degree to which individual measurements vary from the average measurement. In terms of your carbon dioxide data measurements, the variance is the degree to which the individual lab team measurements vary from the average measurement of all of the lab teams together.

Part A:

What follows is a step-by-step guide in determining the variance of a hypothetical set of carbon dioxide measurements:

1. Set up the following four-column data table:

Lab Team #	Milliliters of Carbon Dioxide Produced over a 24-Hour Period	The Difference between the Average and Each Individual Score	The Differences Squared
1	250		
2	60		
3	100		
4	175		
5	40		
6	340		
7	70		
8	110		
9	225		
10	60		

2. Determine the average number of ml of carbon dioxide produced by the 10 lab teams by adding the second column of figures and dividing the sum by the number of measurements in the column. THE AVERAGE = _____.

3. The third column is filled in by determining the difference between the average and each individual measurement.

4. The fourth column is filled in by squaring each number from the third column.

5. Determine the sum of the numbers in the fourth column. THE SUM = _____.

6. The final step in calculating the variance is to use the following formula:

$$\text{The Variance} = \frac{\text{The Sum of the Fourth Column}}{\text{The Total Number of Measurements} - 1}$$

The Variance = _____.

(NOTE: A variance of 0 indicates that there was no variance between the lab teams. In theory, this means that each lab team had collected identical amounts of carbon dioxide. Examining the second column above, one can see that this was not the case.

The larger the variance number, the greater the data varied between the lab teams.)

Part B:

Now you are going to calculate the variance from the data collected during your first investigation.

The data table has been set up for you as follows:

Lab Team #	Milliliters of Carbon Dioxide Produced over a 24-Hour Period	The Difference between the Average and Each Individual Score	The Differences Squared
1			
2			
3			
4			
5			
6			
7			
8			
9			
10			

The calculation of the variance using the formula:

The variance for the first investigation = _____.

Part C:

Now calculate the variance from the data collected during your second investigation.

The variance for the second investigation = _____.

Was the variance decreased as a result of your second investigation? _____

E.   Students Carry Out Their Revised Experimental Procedure in an Attempt to Reduce the Variance

From this point on, you and your students will have to decide how long to keep trying to lower the variance. This can be quite a challenge. However, by carrying out investigations and modifying experimental procedures, significant reduction in the variance can be achieved.

4.7   A FINAL S-A-S UNIT

Using the reagent indophenol (2,6-dichloroindophenol), this investigation involves analyzing foods such as fruits and fruit juices as well as vitamin C tablets for relative amounts of vitamin C.

A.   The Student Seminar Introduces the Problem and the Individual Lab Teams Design Investigations

Explain to students that they are going to design experimental procedures to investigate the relative amounts of vitamin C (ascorbic acid) contained in various fruits, juices, and vitamin C tablets.

Perform the following demonstration for the class:

- Pour 1 inch of a 0.1% solution of indophenol into a test tube.
- Add ascorbic acid solution drop by drop to the test tube, gently agitating the test tube after each drop.
- Count the number of drops of ascorbic acid needed to decolorize the indophenol solution.
- Repeat the preceding demonstration, but this time with a solution of ascorbic acid that has been diluted with water to 50%.

Engage students in discussing what they learned from the demonstration that they can use in their upcoming investigation. (The answer is as follows: One can use indophenol as an indicator of the relative amounts of vitamin C in solutions. The fewer drops of a solution that it takes to decolorize the indophenol, the less the amount of vitamin C that solution contains.)

You might wish to have some lab groups investigate fruit juices, some fruits, and some vitamin C tablets. In any event, the next task is to design the experimental procedures, including data tables.

Worksheet S-A-S 11 is not for student distribution. It presents some experimental procedures that have been used in investigating vitamin C.

B.   Students Carry Out Their Investigations

After students have carried out their investigations, conduct a seminar discussion of the findings.

Examples of Experimental Procedures and Data Tables Used in Investigating the Relative Amounts of Vitamin C in Fruit Juices, Fruits, and Vitamin C Tablets

A. For testing each fruit juice selected:

1. Carefully pour 10 ml of 0.1% indophenol solution into a clean test tube.
2. Drop by drop, add the fruit juice being tested. Gently agitate the test tube after each drop.
3. Continue adding drops of the fruit juice until the indophenol solution is decolorized.
4. Record the number of drops in the following data table.

Kind of Fruit Juice	Brand Name	Number of Drops Added to Decolorize Indophenol
Orange (fresh)		
Orange (frozen)		
Grapefruit (fresh)		
Grapefruit (frozen)		
etc.		

B. For testing each fruit selected:

1. Obtain juice from the selected fruit. Squeezing by hand into a clean beaker is a good method for oranges and grapefruits. For other fruits, such as cherries, wrap them in layers of cheesecloth and then gather up the ends of the cheesecloth and squeeze into a clean beaker.
2. Carefully pour 10 ml of 0.1% indophenol solution into a test tube.
3. Drop by drop, add the juice being tested, gently agitating the test tube after each drop.
4. Continue adding drops of the juice until the indophenol solution is decolorized.
5. Record the number of drops in the following data table.

Name of Fruit	Number of Drops Added to Decolorize Indophenol
Cherries	
Grapes	
Orange	
Pineapple	
Apple	
etc.	

C. For testing each vitamin C tablet selected:

1. Using a mortar and pestle, thoroughly grind up the vitamin C tablet in 10 ml of water.
2. Carefully pour 10 ml of 0.01 indophenol solution into a clean test tube.
3. Drop by drop, add the liquid from the mortar, gently agitating the test tube after each drop.
4. Continue adding drops of the liquid until the indophenol solution is decolorized.
5. Record the number of drops in the following data table.

Brand Name	Size of Tablet (mg)	Number of Drops Added to Decolorize Indophenol
	1000 mg	
	500 mg	
	100 mg	
	etc.	

D. Optional data tables:

Comparing Previously Packaged to Freshly Squeezed
(for each juice compared, circle the number
of drops which represents the juice
with the greater amount of vitamin C)

Name of Juice	# of Drops Added of Previously Packaged	# of Drops Added of Freshly Squeezed
Orange	10	7
Lemon		
Grapefruit		
Apple		

Comparing Brand Names of Previously Packaged Juices
(for each brand name comparison, circle the
brand with the greater amount of vitamin C)

Type of Juice	Brand Name	# of Drops Added	COMPARED TO	Brand Name	# of Drops Added
Orange	Mr. Orange	12		Tops	7

SECTION 5

Sponge Activities

This last section of the Biology Teacher's Survival Guide provides 100 reproducible student worksheet activities you can use at any time. The worksheets are varied in format and cover a wide range of biology topics.

For quick location of appropriate activities, the Contents lists the worksheets in alphabetical order by topic, from "Algae: Word Scramble" (Worksheet 5-1) through "Vitamins and Minerals: Fill in the Blanks" (Worksheet 5-100).

Answers to all of the Sponge Activities follow Worksheet 5-100.

WORKSHEET 5-1

Algae: Word Scramble

Unscramble the underlined word in each sentence and write the correct word on the dotted line.

1. Based on pigments produced by the algae, one group is called the <u>WORNB</u> ____ ____ ____ ____ ____ algae.

           8   3 & 13

2. The classification group under which one finds green algae is <u>PYCHOTAOLHR</u>
    ___ ___ ___ ___ ___ ___ ___ ___ ___ ___ ___.

 1   18

3. An example of a colonial, motile green alga is <u>XOOVLV</u> ___ ___ ___ ___ ___ ___.

4. An example of a filamentous green alga is <u>LTRXIOUH</u> ___ ___ ___ ___ ___ ___ ___.

                                  16

5. An example of a nonmotile, unicellular green alga is <u>OELCLHRLA</u> ___ ___ ___ ___
    ___ ___ ___ ___ ___.

    20   10

6. An example of a filamentous green alga with spiral chloroplasts is <u>ROGSIAYRP</u> ____
    ___ ___ ___ ___ ___ ___ ___.

      17   4   21

7. An example of a unicellular green alga that displays many beautiful shapes and designs
    is the <u>SIMDED</u> ____ ____ ____ ____ ____ ____.

                     12

8. The classification group for brown algae is <u>TAAPPHEOYH</u> ____ ___ ___ ____ ____

                                     15

____ ____ ____ ____ ____.

           6

9. An example of a very common brown alga is <u>ELKP</u> ____ ____ ____ ____.

                                    19

10. The classification group for red algae is <u>HRPODHOTYA</u> ____ ___ ___ ___ ___ ___

                                     2

____ ____ ____ ____ ____.

           5

11. An example of unicellular golden algae that display many different shapes and designs is
    the <u>IADOTMS</u> ____ ____ ____ ____ ____ ____ ____.

              7

12. The pigment in algae that allows for photosynthesis is <u>HLROYLLHPOC</u> ____ ____ ____

                                         14

____ ____ ____ ____ ____ ____.

13. YPPNTOHAKNOTL $\underset{11}{\rule{1em}{0.4pt}}$ $\rule{1em}{0.4pt}$ $\rule{1em}{0.4pt}$ $\rule{1em}{0.4pt}$ $\rule{1em}{0.4pt}$ $\rule{1em}{0.4pt}$ $\rule{1em}{0.4pt}$ $\rule{1em}{0.4pt}$ $\rule{1em}{0.4pt}$ $\rule{1em}{0.4pt}$ $\underset{9}{\rule{1em}{0.4pt}}$

is the name given to algae that comprise plankton.

To answer the following, select the appropriate numbered letters above and write them on the dotted lines to complete each statement.

A. Some algae reproduce by a process called $\underset{1}{\rule{1em}{0.4pt}}$ $\underset{2}{\rule{1em}{0.4pt}}$ $\underset{3}{\rule{1em}{0.4pt}}$ J U $\underset{4}{\rule{1em}{0.4pt}}$ $\underset{5}{\rule{1em}{0.4pt}}$ $\underset{6}{\rule{1em}{0.4pt}}$ $\underset{7}{\rule{1em}{0.4pt}}$

$\underset{8}{\rule{1em}{0.4pt}}$ $\underset{9}{\rule{1em}{0.4pt}}$

B. Two food thickeners obtained from algae are $\underset{10}{\rule{1em}{0.4pt}}$ $\underset{11}{\rule{1em}{0.4pt}}$ G $\underset{12}{\rule{1em}{0.4pt}}$ $\underset{13}{\rule{1em}{0.4pt}}$ and $\underset{14}{\rule{1em}{0.4pt}}$ $\underset{15}{\rule{1em}{0.4pt}}$ $\underset{16}{\rule{1em}{0.4pt}}$

$\underset{17}{\rule{1em}{0.4pt}}$ $\underset{18}{\rule{1em}{0.4pt}}$ G $\underset{19}{\rule{1em}{0.4pt}}$ $\underset{20}{\rule{1em}{0.4pt}}$ N $\underset{21}{\rule{1em}{0.4pt}}$ N.

WORKSHEET 5-2

Amphibians: Fill in the Blanks

Fill in the blanks of the statements below using proper terms selected from the list that follows the statements.

1. Most amphibians live in or near _____.
2. In terms of diet, most amphibians are _____.
3. Amphibians belong to phylum _____.
4. Since amphibians do not maintain a constant body temperature, they are said to be _____.
5. The amphibian heart has _____ chambers.
6. Fertilization of most amphibians is _____.
7. _____ is the process by which amphibians have a complete change of body form as they mature.
8. The term amphibian derives from a word meaning _____.
9. The toads and the _____ are the most abundant amphibians.
10. Unlike some other amphibians in terms of body structure, toads and frogs do not have a _____.
11. The genus name for toads is _____.
12. _____ are a type of amphibian that have tails.
13. A tadpole is the _____ stage of an adult frog.
14. Amphibian eggs have a _____ coating to protect them from drying out.
15. Unlike most frogs, which spend a great deal of time near water, _____ live mostly on land.
16. Young amphibians use _____ while adults use _____ to breathe.
17. _____ is the scientific name for the common grass frog.
18. Unlike reptiles, the skin of amphibians does not contain _____.
19. Unlike amphibian feet, reptile feet contain _____.

frogs	Chordata	water	double life	metamorphosis
larval	carnivores	three	salamanders	Rana pipiens
claws	Bufo	external	toads	tail
cold-blooded	jelly-like	scales	gills	lungs

WORKSHEET 5-3

Animal Development: Matching

For each of the following groups of terms, connect each term in the left-hand column with its proper counterpart at the right by drawing a straight line between them.

Group 1:

a.	larva	a.	Fertilized egg
b.	amniote	b.	Stage of developing insect
c.	zygote	c.	Ripe egg being released from ovary
d.	ovulation	d.	Organ in which mammal embryo develops
e.	ovoviviparous	e.	Embryo develops in egg in mother's body
f.	uterus	f.	Type of bird and reptile eggs

Group 2:

a.	embryology	a.	Ball of cells following zygote
b.	blastocyst	b.	Larval stage of butterfly
c.	caterpillar	c.	Embryo develops inside of mother, but not in egg
d.	vivparous	d.	An embryo over three months old
e.	fetus	e.	Female egg cell
f.	ovum	f.	The study of embryo development

Group 3:

a.	metamorphosis	a.	Final stage of metamorphosis
b.	pupa	b.	Barrier between mother and embryo
c.	gestation	c.	Change from tadpole to frog
d.	adult	d.	Inactive stage of insect following larva
e.	placenta	e.	Structure of an individual
f.	morphology	f.	Time of mammal development in mother

WORKSHEET 5-4

Animal Groups: Matching

Match the following six animal groups with the 60 animals listed. Write the correct animal group letters in the blanks.

B = birds   F = Fish   M = mammals   A = amphibians   R = reptiles   I = insects

_____ 1. camel	_____ 21. opossum	_____ 41. flea
_____ 2. frog	_____ 22. dolphin	_____ 42. praying mantis
_____ 3. butterfly	_____ 23. alligator	_____ 43. cricket
_____ 4. snake	_____ 24. lizard	_____ 44. katydid
_____ 5. bass	_____ 25. mouse	_____ 45. Canada goose
_____ 6. whale	_____ 26. condor	_____ 46. porpoise
_____ 7. raccoon	_____ 27. grasshopper	_____ 47. owl
_____ 8. kangaroo	_____ 28. grouse	_____ 48. moth
_____ 9. seahorse	_____ 29. elephant	_____ 49. weasel
_____ 10. toad	_____ 30. albatross	_____ 50. pheasant
_____ 11. moth	_____ 31. human	_____ 51. cockroach
_____ 12. bat	_____ 32. platypus	_____ 52. termite
_____ 13. turtle	_____ 33. Koala bear	_____ 53. sea lion
_____ 14. seal	_____ 34. shark	_____ 54. mosquito
_____ 15. salamander	_____ 35. lungfish	_____ 55. fly
_____ 16. iguana	_____ 36. lemur	_____ 56. beetle
_____ 17. walrus	_____ 37. monkey	_____ 57. wasp
_____ 18. dinosaur (extinct)	_____ 38. rat	_____ 58. eel
_____ 19. crocodile	_____ 39. ape	_____ 59. stingray
_____ 20. lion	_____ 40. bee	_____ 60. chameleon

WORKSHEET 5-5

Animal Phyla: Name That Category

Read the hints and determine the general category to which they belong. Write the category in the appropriate square. Answers can be found in the list at the bottom of the worksheet.

1. Body segmented into head, thorax, abdomen.
2. Aquatic, body canals and pores.
3. Dorsal nerve cord, some members have a backbone.
4. Radial symmetry, tube feet, water vascular system.
5. Segmented body plan, closed circulatory system.

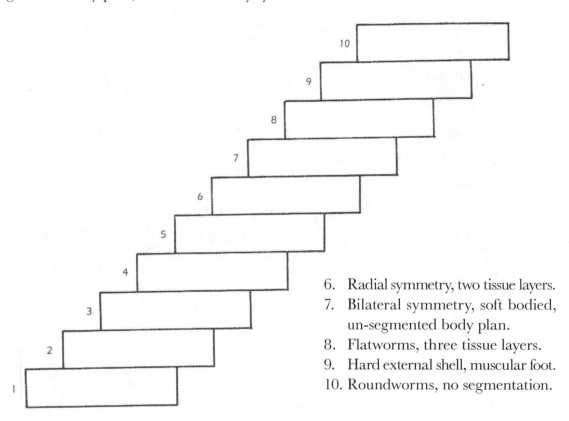

6. Radial symmetry, two tissue layers.
7. Bilateral symmetry, soft bodied, un-segmented body plan.
8. Flatworms, three tissue layers.
9. Hard external shell, muscular foot.
10. Roundworms, no segmentation.

Answers:

Platyhelminthes	Sarcodina	Cnidaria	Arthropoda
Ascomycota	Aschelminthes	Mollusca	Porifera
Echinodermata	Hemichordata	Annelide	Chordata

WORKSHEET 5-6

Animal Phyla Members: Matching

Match the following 10 animal phyla with the 51 animals listed. Write the correct phyla letters in the blanks.

a. Hemichordata   e. Porifera   h. Aschelminthes
b. Echinodermata   f. Cnidaria   i. Platyhelminthes
c. Arthropoda   g. Annelida   j. Mollusca
d. Chordata

_____ 1. sponge	_____ 18. mammals	_____ 35. insects
_____ 2. alligator	_____ 19. nematode	_____ 36. chiton
_____ 3. roundworm	_____ 20. leech	_____ 37. crocodile
_____ 4. sandworm	_____ 21. octopus	_____ 38. millipede
_____ 5. liverfluke	_____ 22. fish	_____ 39. human
_____ 6. squid	_____ 23. tapeworm	_____ 40. tick
_____ 7. sea star	_____ 24. kangaroo	_____ 41. clam
_____ 8. jellyfish	_____ 25. brittle star	_____ 42. snail
_____ 9. horseshoe crab	_____ 26. rotifer	_____ 43. Trichina worm
_____ 10. frog	_____ 27. slug	_____ 44. crayfish
_____ 11. pigeon	_____ 28. earthworm	_____ 45. lobster
_____ 12. amphibians	_____ 29. tubeworm	_____ 46. bird
_____ 13. sand dollar	_____ 30. spider	_____ 47. mussel
_____ 14. coral	_____ 31. centipede	_____ 48. planarian
_____ 15. oyster	_____ 32. tongue worm	_____ 49. sea anemone
_____ 16. reptiles	_____ 33. vertebrates	_____ 50. dog
_____ 17. sea urchin	_____ 34. hydra	_____ 51. mole

WORKSHEET 5-7

Facts of Biology: Letter Search

Locate each letter in the letter scramble and write it in the space to the right of the instruction to spell out the answer to the hint.

Hint 1: The largest rodent in the United States

COLUMN FROM LEFT	ROW FROM TOP										
2	1	____	C	B	I	R	X	B	F	Y	O
6	3	____	H	S	A	D	H	G	L	E	Z
2	4	____	J	C	F	K	R	E	I	M	L
4	6	____	Q	A	N	K	T	P	S	O	P
8	1	____	N	J	Q	U	Y	Z	D	E	G
9	6	____	M	V	U	V	W	E	T	L	R

Hint 2: Elephant enlarged   what?   tusks are actually (two words)

COLUMN FROM LEFT	ROW FROM TOP										
9	6	____									
1	3	____									
4	1	____									
3	3	____									
1	6	____									
3	5	____	D	C	E	P	Q	S	R	J	I
7	2	____	H	L	I	M	P	A	N	Z	A
2	1	____	P	O	E	U	R	F	B	D	O
9	1	____	G	T	Y	C	L	H	N	J	F
6	1	____	B	X	I	G	O	P	I	M	K
9	3	____	R	N	W	S	C	E	S	V	U
5	3	____									
4	6	____									

WORKSHEET 5-8

Facts Of Biology: Letter Search

Locate each letter in the letter scramble and write it in the space to the right of the instruction to spell out the answer.

Hint 1: Location of the smallest bones of the body (two-word answer)

COLUMN FROM LEFT	ROW FROM TOP										
2	1	____									
4	5	____									
6	1	____	A	M	C	R	B	D	X	T	W
2	5	____	X	D	M	C	Q	Y	Z	A	C
8	4	____	E	R	H	E	B	M	C	H	Q
1	3	____	W	C	X	O	F	R	O	L	T
4	3	____	C	D	N	I	G	R	N	V	A
8	2	____	D	F	U	Z	O	M	P	T	W
6	4	____									

Hint 2:   Considered the strongest muscle of the body

COLUMN FROM LEFT	ROW FROM TOP										
2	1	____									
5	4	____									
2	6	____	A	M	E	X	Y	R	S	L	X
7	1	____	C	R	O	Q	B	T	A	D	Z
3	3	____	G	M	E	S	W	C	D	E	I
6	2	____	Y	Q	A	X	A	R	B	N	H
8	3	____	O	A	X	C	W	P	T	U	X
7	6	____	E	S	A	G	B	C	R	W	R

WORKSHEET 5-9

Facts of Biology: Letter Search

Locate each letter in the letter scramble and write it in the space to the right of the instruction to spell out the answer to the hint.

Hint 1: The longest cells in the body

COLUMN FROM LEFT	ROW FROM TOP										
2	6	_____									
8	4	_____	X	P	A	G	B	O	Y	Z	B
4	6	_____	C	W	R	D	M	T	O	B	E
9	3	_____	B	B	N	Q	C	H	K	C	R
2	4	_____	W	O	Z	V	S	A	C	E	B
3	3	_____	O	Y	I	Q	O	U	S	W	T
5	4	_____	N	N	R	U	L	X	M	V	N

Hint 2: What you are doing if you are sternutating

COLUMN FROM LEFT	ROW FROM TOP										
2	2	_____									
5	4	_____									
7	1	_____	Z	R	B	S	N	P	C	Y	D
9	6	_____	H	S	C	Z	R	M	E	D	G
2	6	_____	G	O	Q	I	A	G	C	F	G
4	3	_____	E	W	N	J	N	K	O	M	L
6	5	_____	M	E	C	Q	A	N	G	C	B
9	3	_____	B	Z	M	Y	F	H	K	A	L

WORKSHEET 5-10

Facts of Biology: Letter Search

Locate each letter in the letter scramble and write it in the space to the right of the instruction to spell out the answer to the hint.

Hint 1: The master gland of the body

COLUMN FROM LEFT	ROW FROM TOP										
3	1	____									
6	4	____									
1	4	____									
1	1	____	U	A	P	C	R	T	V	R	X
5	2	____	C	W	R	O	I	Z	D	K	M
3	6	____	O	Y	C	K	J	R	S	C	A
7	6	____	T	B	W	N	U	I	V	B	D
8	1	____	C	O	S	R	T	A	Z	C	Y
9	5	____	B	W	T	Y	R	Q	A	Z	O

Hint 2: What you are doing if you are lacrimating

COLUMN FROM LEFT	ROW FROM TOP									
2	3	____	A	X	G	B	H	Z	O	N	Q
5	5	____	G	M	Q	V	D	J	R	X	I
2	4	____	C	H	C	D	M	W	B	E	O
9	2	____	N	Y	G	N	K	L	I	J	S
4	4	____	P	A	P	O	R	E	Q	S	Z
3	4	____	B	L	B	U	F	H	K	T	F

Name: _____   **Date:** _____

Facts of Biology: Letter Search

Locate each letter in the letter scramble and write it in the space to the right of the instruction to spell out the answer to the hint.

Hint 1: To be in a perfect state of health is to be in a perfect state of what?

COLUMN FROM LEFT	ROW FROM TOP										
9	1	____									
5	1	____									
9	5	____									
2	5	____									
5	3	____	T	A	D	F	O	A	S	B	H
2	6	____	T	M	A	Y	C	Z	D	U	X
1	2	____	H	B	A	N	O	E	E	O	I
3	3	____	J	C	B	S	I	K	T	F	V
2	6	____	I	E	L	O	S	M	G	P	M
9	3	____	K	S	R	G	Q	N	H	J	W
5	5	____									

Hint 2: The heaviest bone of the human body

COLUMN FROM LEFT	ROW FROM TOP									
2	1	____	A	F	C	M	Z	O	Y	A	G
4	2	____	C	N	H	E	Q	B	R	Z	X
6	5	____	Z	W	J	A	C	B	E	R	T
2	5	____	B	I	N	K	Q	M	P	S	F
8	3	____	C	U	R	C	L	M	G	O	D
			O	V	A	G	X	H	Z	U	C

WORKSHEET 5-12

Facts of Biology: Letter Search

Locate each letter in the letter scramble and write it in the space to the right of the instruction to spell out the answer to the hint.

Hint 1: Type of fat, tested for in the blood, that might be of diagnostic value in predicting future heart disease

COLUMN FROM LEFT	ROW FROM TOP	
1	1	_____
5	6	_____
5	1	_____
5	2	_____
6	2	_____
4	1	_____
9	6	_____
3	6	_____
3	2	_____
7	6	_____
1	6	_____
9	2	_____

```
T  A  K  Y  I  F  J  G  A
B  L  R  U  G  L  S  X  E
C  E  D  A  Q  V  W  H  O
T  M  D  N  Ĕ  G  C  Z  L
O  P  Y  D  E  F  C  B  M
D  R  E  H  R  I  I  T  C
```

Hint 2: A steroid, tested for in the blood, that might be of diagnostic value in predicting future heart disease

COLUMN FROM LEFT	ROW FROM TOP	
3	1	_____
8	6	_____
8	1	_____
4	6	_____
6	5	_____
2	6	_____
6	1	_____
9	4	_____
6	3	_____
8	3	_____
1	3	_____

```
A  I  C  F  P  T  M  O  A
K  X  L  B  S  H  T  C  O
L  O  B  E  F  R  C  W  C
J  Q  G  R  D  U  N  D  E
L  Z  Y  I  J  E  V  G  Z
H  S  A  L  O  E  V  H  M
```

WORKSHEET 5-13

Facts of Biology: Letter Search

Locate each letter and write it in the space to the right of the instruction to spell out the answer.

Hint 1: The largest animal on Earth today

COLUMN FROM LEFT	ROW FROM TOP									
1	1	____								
4	3	____								
1	5	____	B	I	Z	A	Y	E	A	L	Q
6	1	____	D	F	O	R	W	L	C	L	B
6	3	____	F	H	T	L	C	W	M	Z	R
2	3	____	E	O	U	Z	A	H	V	C	E
7	1	____	U	F	T	B	P	W	K	E	L
4	3	____	S	R	G	I	J	X	Q	D	Y
9	4	____									

Hint 2: Name of tree that can reach a height of 300 feet and an age of 4000 years old

COLUMN FROM LEFT	ROW FROM TOP								
6	6	____								
4	4	____	C	Y	M	T	O	B	S	R	Q
7	2	____	A	S	E	I	Q	T	Q	W	B
3	5	____	Z	A	K	O	O	L	A	Z	C
5	3	____	A	D	U	E	G	R	V	A	B
7	5	____	B	J	U	F	Y	S	I	C	N
1	4	____	R	C	M	P	X	S	B	D	H

WORKSHEET 5-14

Facts of Biology: Letter Search

Locate each letter in the letter scramble and write it in the space to the right of the instruction to spell out the answer to the hint.

Hint 1: The disease that is responsible for more deaths per year than any other

COLUMN FROM LEFT	ROW FROM TOP										
2	1	_____									
3	1	_____									
4	1	_____									
5	2	_____									
6	3	_____									
7	5	_____	A	C	A	R	D	G	T	F	A
1	2	_____	V	M	O	H	D	P	S	R	D
2	3	_____	C	A	X	Y	L	I	A	B	G
7	2	_____	I	N	S	U	C	B	K	B	F
4	5	_____	J	W	K	C	U	E	O	Q	J
4	4	_____	L	R	A	Z	L	D	V	E	C
5	6	_____									
3	6	_____									
2	6	_____									

Hint 2: The largest internal organ of the body

COLUMN FROM LEFT	ROW FROM TOP									
3	1	_____	B	I	L	M	A	E	E	P	D
4	2	_____	F	X	R	I	G	N	H	S	S
3	4	_____	Q	A	W	Y	K	C	D	I	L
7	1	_____	R	E	V	B	U	R	V	N	P
6	4	_____	J	M	F	S	A	G	R	U	Z
		_____	C	O	Q	O	L	H	T	A	K

WORKSHEET 5-15

Facts of Biology: Letter Search

Locate each letter in the letter scramble and write it in the space to the right of the instruction to spell out the answer to the hint.

Hint 1: Another name for a kiss

COLUMN FROM LEFT	ROW FROM TOP										
2	1	____									
4	1										
9	6	____	A	O	R	S	T	Z	C	F	L
3	5	____	O	B	G	N	Q	A	T	R	O
9	1	____	Y	N	C	B	F	W	I	Z	A
2	5	____	D	C	R	O	M	Z	Q	K	B
8	6	____	G	A	U	C	R	A	Z	M	O
7	3	____	M	B	D	X	C	D	B	T	C
4	4	____									
2	3	____									

Hint 2: Another name for a yawn

COLUMN FROM LEFT	ROW FROM TOP										
6	1	____									
9	6	____									
4	6	____	Z	R	S	A	I	O	M	P	Q
6	4	____	B	C	I	E	B	N	O	C	F
7	5	____	Y	N	O	Z	K	B	F	C	H
1	6	____	R	G	R	I	E	L	N	P	J
3	1	____	M	K	A	G	O	Q	A	X	R
5	5	____	T	D	I	U	R	T	M	X	S
2	3	____									

WORKSHEET 5-16

General Biology: Around the Square

Using the clues and answers provided, fill in the squares with the appropriate answers beginning with number 1 at the lower left corner and continuing clockwise. (Selected numbers around the square are provided as a guide.)

Clues:

1. What the term biology means
2. A one-celled organism that contains chloroplasts and moves by means of flagella
3. The system of measurement used by scientists
4. Anton van Leeuwenhoek designed and made many of these valuable laboratory instruments.
5. One of the achievements that this scientist is remembered for is the germ theory of disease.
6. This structure, found in cells, produces proteins.
7. Made of DNA, this structure contains a string of genes.
8. This scientist developed the first classification system.
9. The male reproductive organ of plants
10. The amoeba and the Paramecium belong to this group of one-celled microorganisms.
11. The basic unit of structure of living things

Answers:

Linnaeus	protozoans
ribosome	Metric system
stamen	chromosome
study of life	microscope
cells	Louis Pasteur
Euglena	

Name: _____   Date: _____

WORKSHEET 5-17

General Biology: Name That Category

Read each list of hints and determine the general category to which each list belongs. Write the category in the appropriate square.

1. Tracheophytes, Porifera, Echinodermata, Chordata
2. Student, binocular, electron
3. Carbon dioxide, oxygen, glucose, water
4. Dominance, independent assortment, segregation
5. Spirillum, bacillus, coccus
6. Pseudopods, flagella, cilia
7. Mushrooms, molds, yeasts

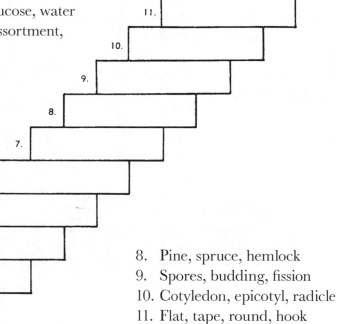

8. Pine, spruce, hemlock
9. Spores, budding, fission
10. Cotyledon, epicotyl, radicle
11. Flat, tape, round, hook
12. Head, thorax, abdomen

Categories:

Koch and genetics	seed parts	phylum names	deciduous
earthworm	microorganism types	kingdom names	algae
protozoan types	Mendel and genetics	bacteria shapes	fungi
class names	conifers	stem parts	insects
sexual reproduction	photosynthesis	mitosis stages	worms
asexual reproduction	locomotion structures	microscope types	meiosis

Name: _____   Date: _____

WORKSHEET 5-18

General Biology: Name That Category

Read each list of hints and determine the general category to which each list belongs. Write the category in the appropriate square.

1. Bread-cereal, meat, dairy, fruit-vegetable
2. Adenine, cytosine, guanine, thymine, sugar, phosphate
3. Prophase, anaphase, metaphase, telophase
4. Nucleus, ribosomes, mitochondria, vacuole
5. Humerus, femur, radius, tibia, vertebra
6. Stage, revolving nosepiece, ocular, arm
7. A, B₁, B₂, C, D, K

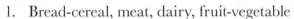

8. Volvox, Spirogyra, Ulothrix
9. Monera, Protista, animal, plant, fungi
10. Osmosis, diffusion, pinocytosis, phagocytosis
11. Liverworts, mosses, Sphagnum moss
12. Pistil, stamen, sepals, petals

Categories:

phyla	flower parts	brown algae	minerals
vitamins	kingdoms	vascular plants	skeletal bones
algae parts	mitosis stages	green algae	amino acids
red algae	nonvascular plants	food groups	DNA parts
cell parts	types of cell division	microscope parts	classes
cell transport	trees	bacteria	orders

Name: _____   **Date:** _____

General Biology: True and False

Read each statement carefully. If the underlined word makes the statement true, write a T in the blank; if false, write an F. Correct the false statements by crossing out the underlined word and writing the proper word above it.

_____ 1. The food chain begins with organisms known as <u>consumers</u>.

_____ 2. The three basic categories of food are fats, carbohydrates, and <u>lipids</u>.

_____ 3. <u>Diffusion</u> is restricted to the passing of water molecules through a membrane from a region of higher concentration to a region of lower concentration.

_____ 4. ATP compounds in the cell are energy <u>rich</u>.

_____ 5. <u>Louis Pasteur</u> is considered the Father of Genetics.

_____ 6. Evidence suggests that humans probably originated in <u>South America</u>.

_____ 7. The Kingdom Monera contains organisms such as <u>molds</u>.

_____ 8. All organisms are given a name consisting of species and <u>genus</u>.

_____ 9. Clams and snails are <u>Mollusks</u>.

_____ 10. Mushrooms <u>are</u> fungus organisms.

_____ 11. An organism that can make its own food is known as a <u>heterotroph</u>.

_____ 12. Plants take in <u>carbon dioxide</u> from the air for photosynthesis.

_____ 13. Part of the cell theory states: <u>most</u> cells are produced from other cells.

_____ 14. <u>Ribosomes</u> are sites of protein synthesis within the cell.

_____ 15. <u>Mitosis</u> is division of the nucleus of a cell.

_____ 16. The ability of an organism to survive in an ever-changing environment is called <u>adaptation</u>.

_____ 17. The scientific name for humans is <u>Homo erectus</u>.

_____ 18. The Phylum <u>Tracheophyta</u> contains plants with true roots, stems, and leaves.

_____ 19. <u>Translocation</u> is the process by which water evaporates from a plant.

_____ 20. All the regions of the Earth where life can be found are called the <u>lithosphere</u>.

WORKSHEET 5-20

Biomes: Matching

Select the proper terms listed in the right-hand column to match the statements in the left-hand column. Write the letter of the term in the blank.

_____	1.	A chart or graph containing monthly precipitation and temperature data for an area	a.	marine
_____	2.	Regions of the Earth characterized by distinct plants and animals	b.	terrestrial
_____	3.	A biome in which grasses predominate	c.	tundra
_____	4.	A scientist who studies the relationship between organisms and their environments	d.	grassland
_____	5.	A collective term for all of the ocean biomes	e.	desert
_____	6.	Comprised of deciduous trees, this biome experiences four definite seasonal changes annually.	f.	temperate deciduous forest
_____	7.	A unit of geographical distance on a map north and south of the equator	g.	biome
_____	8.	Found at the northernmost latitudes, this biome receives precipitation mostly in the form of snow.	h.	coniferous forest
_____	9.	The tropical rain forest biome	i.	climatogram
_____	10.	A collective term for all of the land biomes	j.	ecologist
_____	11.	The tundra biome	k.	latitude
_____	12.	Made up primarily of cone-bearing trees, this biome is located just south of the tundra.	l.	tropical rain forest
_____	13.	Located in the equatorial regions, this biome experiences heavy rainfall throughout the year and a constant very warm temperature.	m.	biome that receives most annual rainfall
_____	14.	This biome receives less than 25 cm of rain per year, has a high evaporation rate, and can experience a wide range of temperatures.	n.	biome that receives least annual rainfall
_____	15.	The temperate deciduous forest biome	o.	biome that comprises the eastern United States

WORKSHEET 5-21

Body Movements and Positions: Name That Category

Read the hints and determine the proper category to which they belong. Write the category in the appropriate box. Categories can be found in the list at the bottom of the worksheet.

1. Moving an arm or a leg straight out to the side
2. Turning a bone on its own axis such as turning the head
3. Bending the arm as in "making a biceps muscle"
4. Turning the hand so the palm faces down
5. The body standing upright with palms facing forward

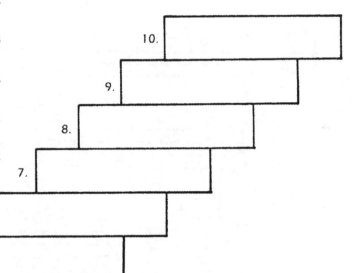

6. The body reclined on its back, face up
7. Straightening the forearm after having bent it
8. Turning the hand so that the palm faces up
9. Bringing an arm or leg back into position after having moved it out to the sides of the body
10. The body reclined on its stomach, face down

Categories:

pronation	flexion	abduction
extension	adduction	supinated
anatomical position	pronated position	supination
rotation	circumduction	retraction

WORKSHEET 5-22

Bones of the Body: Fill in the Blanks

Read the hints and select the correct answers from the terms at the bottom of this worksheet. Write each term in the proper blank.

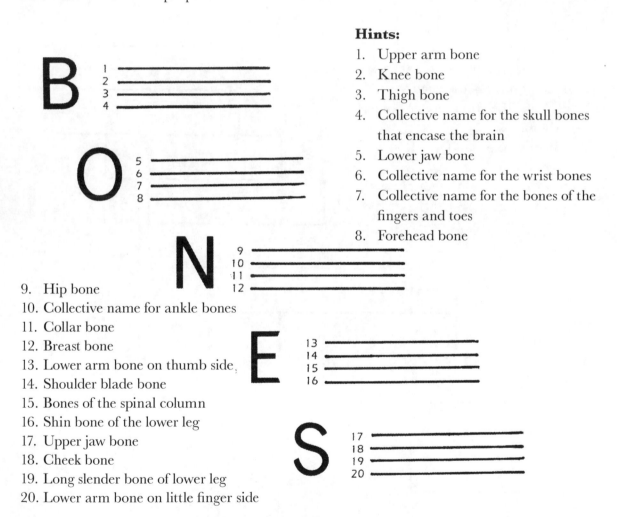

Hints:
1. Upper arm bone
2. Knee bone
3. Thigh bone
4. Collective name for the skull bones that encase the brain
5. Lower jaw bone
6. Collective name for the wrist bones
7. Collective name for the bones of the fingers and toes
8. Forehead bone
9. Hip bone
10. Collective name for ankle bones
11. Collar bone
12. Breast bone
13. Lower arm bone on thumb side
14. Shoulder blade bone
15. Bones of the spinal column
16. Shin bone of the lower leg
17. Upper jaw bone
18. Cheek bone
19. Long slender bone of lower leg
20. Lower arm bone on little finger side

Terms:

innominate	tibia	humerus	mandible	malar (zygomatic)
clavicle	femur	frontal	radius	vertebrae
ulna	patella	scapula	cranium	maxilla
carpals	fibula	sternum	phalanges	tarsals

Name: _____ Date: _____

WORKSHEET 5-23

Brain and Brain Stem Functions: Matching

Select the correct term from the right-hand column to match the proper statement from the left-hand column. Write the letter of the terms in the blanks provided. (Some terms will be used more than once.)

_____ 1. The lobe that deals with higher thought processes such as learning from past experiences and planning ahead

_____ 2. It contains vital reflex centers for blood pressure, respiration, and heart rate.

a. cranial nerves

_____ 3. The outer, thin layer of the cerebrum where "the work of the cerebrum" is accomplished

b. medulla

_____ 4. The lobe that has a center that deals with taste

c. frontal

_____ 5. The fissure that separates the temporal from the parietal lobe

d. Rolando

_____ 6. The lobe that deals with vision

e. midbrain

_____ 7. It contains centers for pupillary reflexes and eye movement reflexes.

f. medulla oblongata

_____ 8. It deals with coordinating muscle movement.

g. cerebellum

_____ 9. The lobe that receives sensory information from skin receptors

h. occipital

_____ 10. The lobe that deals with the control of voluntary muscle movements

i. pons

_____ 11. The inner white matter of the cerebrum that functions in communication between the lobes

j. Sylvius

_____ 12. The lobe that deals with smell and hearing

k. parietal

_____ 13. It controls reflexes such as coughing, sneezing, and swallowing.

l. cortex

_____ 14. The fissure that separates the parietal from the frontal lobe

m. temporal

_____ 15. It contains a center that helps with the control of respiration.

_____ 16. There are 12 pairs of these that branch out from the brain.

WORKSHEET 5-24

The Brain, Brain Stem, and Spinal Cord: Color Coding

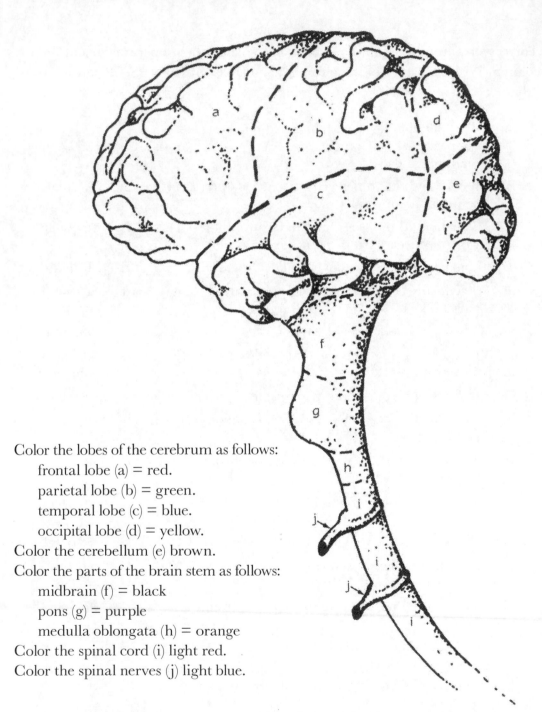

Color the lobes of the cerebrum as follows:

    frontal lobe (a) = red.

    parietal lobe (b) = green.

    temporal lobe (c) = blue.

    occipital lobe (d) = yellow.

Color the cerebellum (e) brown.

Color the parts of the brain stem as follows:

    midbrain (f) = black

    pons (g) = purple

    medulla oblongata (h) = orange

Color the spinal cord (i) light red.

Color the spinal nerves (j) light blue.

WORKSHEET 5-25

Bryophytes: Matching

Select the correct term from the right-hand column to match the proper statement from the left-hand column. Write the letter of the term in the blanks provided.

_____ 1. The type of environment in which bryophytes thrive   a.   archegonium

_____ 2. Bryophytes are considered as a group to be what   b.   cutin
        kind of plants?

_____ 3. These two organisms are bryophytes.   c.   phylum

_____ 4. Bryophytes do not have these true structures.   d.   moist

_____ 5. Bryophytes can carry on photosynthesis because   e.   nonvascular
        they contain this material.

_____ 6. These structures anchor bryophytes to the soil.   f.   water

_____ 7. Bryophytes do not have this type of tissue, which   g.   spore capsule
        characterizes more complex plants.

_____ 8. Many bryophytes have a coating of this material,   h.   primitive
        which retards water loss by evaporation.

_____ 9. During the sexual stage of reproduction of   i.   Sphagnum
        bryophytes, this structure is the result of an egg
        being fertilized by a sperm.

_____ 10. Bryophytes have this type of tissue.   j.   mosses and
                                                  liverworts

_____ 11. The name of a liverwort   k.   stems, leaves,
                                                  and roots

_____ 12. Sperm is formed in this structure.   l.   chlorophyll

_____ 13. In order for sperm to reach eggs, this must be   m.   antheridium
        present.

_____ 14. During the asexual stage of reproduction of   n.   rhizoids
        bryophytes, this structure is formed.

_____ 15. Eggs are produced in this structure.   o.   zygote

_____ 16. The name of the moss that is sold as peat moss for   p.   Marchantia
        conditioning the soil

_____ 17. Bryophyta is the name of this classification   q.   vascular
        category for bryophytes.

WORKSHEET 5-26

Cell Parts and Functions: Fill in the Blanks

Select the proper cell parts listed below to fill in the blanks of the sentences that follow.

nucleus	cell membrane	cell wall	mitochondria
cytoplasm	nuclear membrane	nucleolus	vacuoles
Golgi complex	centriole	lysosomes	ribosomes
Chloroplast	chromatin	endoplasmic reticulum	

1. The _____ is the semiliquid portion of the cell in which the cell parts are located.

2. The _____ is referred to as the headquarters of cell operations.

3. The structure found in plant cells, but not animal cells, that carries out the process of photosynthesis is the _____.

4. Proteins are manufactured by the _____.

5. The structure that surrounds the cell and regulates what enters and leaves the cell is the _____.

6. Nicknamed the "powerhouse of the cell," the _____ are involved in energy production for the cell.

7. The _____ surrounds the nucleus and controls what enters and leaves it.

8. _____ are structures that contain digestive enzymes.

9. In addition to a cell membrane, plant cells also have a _____ that serves to provide strength and support to the cell.

10. Storage chambers within the cell are called _____.

11. Found mostly in animal cells, the _____ plays a role in cell division.

12. The cell structure that prepares and packages proteins either for use within the cell or for shipment out of the cell is the _____.

13. Located within the nucleus, the _____ is involved in making ribosomes for the cell.

14. Among other tasks, the _____ serves as a transportation system within the cell.

15. Another name for the DNA material located within the nucleus of the cell is _____.

WORKSHEET 5-27

Cell Structure: Color Coding

Color the numbered parts of the cell below using the color coding guide at the bottom of the sheet.

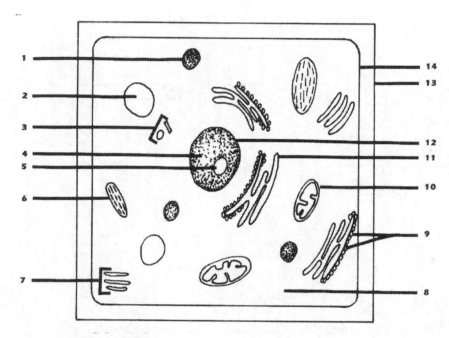

Color Coding Guide:

1. Lysosomes = light red
2. Vacuoles = leave white
3. Centrioles (found mostly in animal cells) = purple
4. Chromatin (DNA material) = dark red
5. Nucleolus = black
6. Chloroplasts (not found in animal cells) = green
7. Golgi complex = dark blue
8. Cytoplasm = yellow stipples
9. Ribosomes = black
10. Mitochondria = yellow
11. Endoplasmic reticulum = light blue
12. Nuclear membrane = brown
13. Cell wall (not found in animal cells) = orange
14. Cell membrane = brown

WORSHEET 5-28

Basic Chemistry: Message Square

Cover the left-hand column with a 3 × 5 card. Reveal one letter at a time, and write it in the proper square(s). Write the anticipated answer in the right column.

Hint: It is the type of event that occurs when two or more materials are combined, resulting in a new substance.

Anticipated Answer

Put: an A in 7 and 11.
      an L in 8.
      an I in 5 and 14.
      an N in 16.
      an H in 2.
      an R in 9.
      an E in 3 and 10.
      a C in 1, 6, and 12.
      an M in 4.
      a T in 13.
      an O in 15.

| 1 | 2 | 3 | 4 | 5 | 6 | 7 | 8 | ■ | 9 | 10 | 11 | 12 | 13 | 14 | 15 | 16 |

RESEARCH QUESTIONS FOR TEXT OR LIBRARY   ANSWERS   SOURCE (INCLUDE PAGES)

1. What is the type of change that results when two or more materials are combined that do not result in a new substance?

2. What is the basic difference between a mixture and a compound?

3. What is a solution?

4. In preparing a solution, what is the difference between the solvent and the solute?

5. Is a suspension a solution or a mixture? Explain.

Name: _____   **Date:** _____

WORKSHEET 5-29

Basic Chemistry: Message Square

Cover the left-hand column with a 3 × 5 card. Reveal one letter at a time, and write it in the proper square(s). Write the anticipated answer in the right column.

Hint: It is comprised of symbols and numbers that indicate the kind and number of atoms that make up a compound.

Put: an M in 4 and 12.
          an L in 8 and 14.
          a C in 1 and 6.
          an H in 2.
          an A in 7 and 15.
          an F in 9.
          a U in 13.
          an E in 3.
          an R in 11.
          an O in 10.
          an I in 5.

Anticipated Answer

1 2 3 4 5 6 7 8 ■ 9 10 11 12 13 14 15

RESEARCH QUESTIONS FOR TEXT OR LIBRARY

ANSWERS   SOURCE (INCLUDE PAGES)

1. What is an atom?

2. What is a molecule?

3. What is an element?

4. How many elements have been discovered?

5. What is an ion?

_____

_____

6. In terms of ions, what is the difference between an acid and a base?

_____

_____

_____

_____

WORKSHEET 5-30

The Circulatory System: Sentence Correction

Each sentence has a boldfaced word or phrase that does not belong in the sentence but does belong in one of the other sentences. Correct each sentence by writing in the proper boldfaced word.

1. Red blood cells are called **leukocytes.**

2. The waxy-like substance that can build up along the walls of arteries is **plasma**.

3. The volume percent of red blood cells is called the **differential count**.

4. The function of erythrocytes is to **protect the body against disease**.

5. The **Vena Cava** is a vessel carrying blood leaving the heart for all parts of the body except the lungs.

6. A blood clot is called a **myocardial infarction**.

7. Another name for platelet blood cells is **erythrocytes**.

8. The alternate expansion and recoil of an artery is the **sinoatrial node**.

9. White blood cells are called **thrombocytes**.

10. The liquid part of the blood minus the blood cells is called **cholesterol**.

11. The function of thrombocytes is to **carry food and oxygen**.

12. The function of leukocytes is to **clot blood**.

13. The **pulmonary** is a vessel carrying blood into the heart from all parts of the body except the lungs.

14. Arteries have **valves** to help keep the blood flowing between heart contractions.

15. A blood cell count which determines the percent of each kind of white blood cell is called a **hematocrit**.

16. Veins have **elastic walls** to prevent blood from flowing backward.

17. A heart attack is called a **thrombus**.

18. The **aorta** is a vessel carrying blood from the heart to the lungs.

19. The pace maker of the heart is called the **electrocardiogram**.

20. A visual record of heart contractions is the **pulse**.

WORKSHEET 5-31

Classification of Monerans, Protists, and Fungi: Matching

For each of the following groups of terms, connect each term in the left-hand column with its proper counterpart at the right by drawing a straight line between them.

Group 1:

a. Chlorophyta
b. Bryophyta
c. Tracheophyta
d. Phaeophyta
e. Rhodophyta

a. Mosses
b. Vascular plants
c. Brown algae
d. Red algae
e. Green algae

Group 2:

a. Euglenophyta
b. Sporozoa
c. Mastigophera
d. Chrysophyta
e. Sarcodina
f. Ciliata

a. Plasmodium
b. Trypanosoma
c. Amoeba
d. Paramecium
e. Euglena
f. Golden algae

Group 3:

a. Schizomycetes
b. Cyanophyta
c. Zygomycota
d. Basidiomycota
e. Sarcodina
f. Ascomycota

a. Move using pseudopodia
b. Bread mold
c. Club fungi
d. Bacteria
e. Baker's yeast
f. Blue-green algae

WORKSHEET 5-32

Classification: Message Square

Cover the left column with a 3 × 5 card. Reveal one letter at a time, and write it in the proper square(s). Write the anticipated answer in the right column.

Hint: Three kingdoms under which organisms are classified

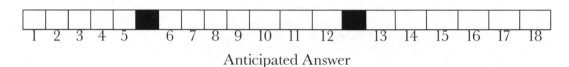

Anticipated Answer

Put: an N in 4 and 14.
      an I in 10 and 15.
      an S in 11.
      an L in 2 and 18.
      an O in 8.
      an A in 3, 13, and 17.
      a P in 1 and 6.
      an M in 16.
      a T in 5, 9, and 12.
      an R in 7.

RESEARCH QUESTIONS FOR TEXT OR LIBRARY	ANSWERS	SOURCE (INCLUDE PAGES)
1. Why was the second kingdom above developed?		
2. Name two additional kingdoms that some classification systems use. Include at least one group of organisms classified under each.		
3. Name the Swedish scientist who developed the system of classification that serves as a model for today's system of classification.		

WORKSHEET 5-33

Classification: Crossword

Match the clues with the answers to complete the crossword.   **ANSWERS:**

CLUES:

Across:
2. Cells whose organelles have a membrane
4. Last word of scientific name
6. Study of classification
7. Developed the system of classification
9. The basic language of classification
10. Classification level between class and family
12. Organisms classified under phylum Arthropoda
14. Broadest groups of classification
15. Protozoans belong to this kingdom

Down:
1. Classification naming system (two words)
5. First term in a scientific name
8. Making its own food
15. Term for molds, mildews, and mushrooms
16. Unable to make its own food

Up:
3. The scientific name for the dog (two words)
11. The scientific name for the house cat (two words)
13. Cells whose nucleus and other organelles do not have a membrane

heterotrophic
taxonomy
Canis familiaris
autotrophic
fungi
eukaryotic
genus
species
Linnaeus
Protista
binomial nomenclature
Latin
Felis domesticus
prokaryotic
kingdom
insects
order

Name: _____ **Date:** _____

WORKSHEET 5-34

The Cranial Nerves: Fill in the Blanks

Select the proper cranial nerve from the list at the bottom of the worksheet and write its name and number in the blank below the following functions. (*Note:* Each cranial nerve is used once.)

vision	face and head sensations	eyeball movements
_____	_____	_____

movements of the tongue	smell	taste saliva secretion facial muscle movements
_____	_____	_____

eyeball movements pupil size regulation	turning head	eyeball movements
_____	_____	_____

taste perception swallowing	hearing balance	slows heart rate movements of internal organs such as stomach and larynx
_____	_____	_____

List of the names and numbers of the 12 cranial nerves:

Olfactory (1)	Trigeminal (5)	Glossopharyngeal (9)
Optic (2)	Abducens (6)	Vagus (10)
Oculomotor (3)	Facial (7)	Accessory (11)
Trochlear (4)	Acoustic (8)	Hypoglossal (12)

WORKSHEET 5-35

Digestion: Word Scramble

Read each clue and then unscramble the words in parentheses, writing the answers in the blanks provided.

1. Daily food intake should include this material in both soluble and insoluble forms.

   (BREIF) __ __ __ __ __

2. One of the final products of fat digestion

   (TTAFY) __ __ __ __ __   (CDAIS) __ __ __ __ __
   ${}_{3}$

3. The final product of protein digestion

   (MAION) __ __ __ __ __   (DCASI) __ __ __ __ __

4. The final product of carbohydrate digestion

   (UGLOCES) __ __ __ __ __ __ __
   ${}_{10}$

5. What polysaccharides are (GSRSUA) __ __ __ __ __ __
   ${}_{8}$

6. They catalyze chemical reactions of digestion.

   (YESNZEM) __ __ __ __ __ __ __

7. Muscular waves that move materials along the digestive tract

   (RPLISITSAES) __ __ __ __ __ __ __ __ __ __ __
   ${}_{4}$

8. Stored in the gall bladder, this material physically breaks down large globules of fat into smaller ones.

   (LEBI) __ __ __ __
   ${}_{9}$

9. The major organ of digestion

   (MSLAL) __ __ __ __ __   (TNISTENEI) __ __ __ __ __ __ __ __ __

10. A group of enzymes that chemically digest fats

    (PILSESA) __ __ __ __ __ __ __
    ${}_{6}$

11. Glands that contain digestive enzymes that are secreted into the mouth

    (AILVAYRS) __ __ __ __ __ __ __ __
    ${}_{7}$

12. The second of the three sections of the small intestine

    (EUJNJMU) __ __ __ __ __ __ __

13. The first section of the large intestine

(SACNENIDG) __ __ __ __ __ __ __ __ __   (OLONC) __ __ __ __ __

14. The name of the structure that carries food from the mouth to the stomach

(SAGEOPUSH) __ __ __ __ __ __ __ __ __
                           1

15. The name for the mixed mass of food and gastric juice in the stomach

(HYEMC) __ __ __ __ __
          2

A.   To learn another name for digestion, select the appropriate numbered letters from above and write them on the dotted lines:

__ __ __ __ __ __ __ __ __ __
1   2   3   4   5   6   7   8   9   10

WORKSHEET 5-36

Organs of the Digestive System: Matching

Select the proper terms listed below to match the statements that follow. Write the letter of the term in the blank.

a. ilium	e. esophagus	i. pancreas	m. tongue	p. peristalsis
b. liver	f. duodenum	j. submandibular	n. teeth	q. stomach
c. colon	g. mouth	k. sublingual	o. jejunum	r. large intestine
d. gallbladder	h. villi	l. parotid		

_____ 1. Stores a substance called bile, which physically breaks down fat droplets.

_____ 2. The third and longest section of the large intestine.

_____ 3. Another name for the large intestine.

_____ 4. Limited digestion begins when salivary gland secretions enter this.

_____ 5. Minute projections extending from the walls of the small intestine and involved in absorbing products of digestion.

_____ 6. Used to cut, tear, and grind food; an adult has 32.

_____ 7. Functions include mixing food and serving as a reservoir prior to the food being passed on to the duodenum.

_____ 8. This pair of salivary glands is located just below the jaw at the back of the mouth.

_____ 9. When you swallow, muscular movements carry the food along this structure to the stomach.

_____ 10. This organ absorbs excess water from undigested food prior to its release from the body as solid waste.

_____ 11. Muscular movements involving the walls of the digestive tract that serve to mix materials and move them along the tract.

_____ 12. This pair of salivary glands is located in the floor of the mouth beneath the tongue.

_____ 13. The short section of the small intestine that leaves the stomach.

_____ 14. It pushes food to the back of the mouth prior to swallowing.

_____ 15. This organ produces bile and detoxifies a variety of substances.

_____ 16. The section of the small intestine between the duodenum and ilium.

_____ 17. Produces digestive juices that are released into the duodenum.

_____ 18. This pair of salivary glands is located just in front of and below the ears.

Name: _____ Date: _____

The Structure of DNA: Color Coding/Fill in the Blanks

Color Coding

Using the instructions in the right-hand column, color the model of DNA sketched below.

DNA MODEL

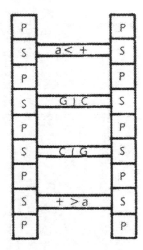

COLOR CODING INSTRUCTIONS

Color: the phosphate units (p) yellow.
the sugar units (s) orange.
the cytosine units (c) red.
the guanine units (g) black.
the adenine units (a) green.
the thymine units (t) blue.

Fill in the Blanks

Select the proper words listed below and fill in the blanks of the sentences that follow.

1. Cytosine, guanine, thymine, and adenine are referred to as _____ bases.
2. Cytosine is always paired with _____ and adenine is always paired with _____.
3. The sides of the DNA molecule are made up of alternating _____ and _____ units.
4. _____ are essentially molecules of DNA.
5. The _____ theory accounts for the method by which a molecule of DNA makes a copy of itself.
6. Based on their own work and that of Maurice Wilkins, _____ and _____ developed a visual model of the DNA molecule.
7. _____ are actually segments along a DNA molecule.

| sugar | zipper | phosphate | guanine | thymine |
| James Watson | nitrogen | genes | chromosomes | Francis Crick |

Name: _____ **Date:** _____

Directional Terms Used in Association with the Body: Crossword

Match the clues to the answers provided to complete the crossword.

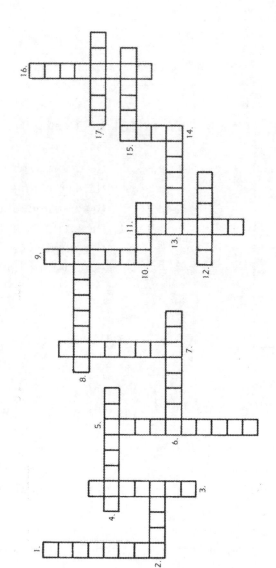

Down:
1. Front surface of the body
5. A horizontal plane or cut separating the body into upper and lower sections
9. A plane or cut separating the body into front and back sections
11. Another name for the anterior surface of the body
16. Closer to the point of attachment to the body

8. Back surface of the body
10. Opposite of hate
12. Another name for the posterior surface of the body
13. Opposite of something
15. Farther away from the point of attachment to the body
17. Inside surface of the arms and legs

Across:

2. Opposite of wrong
4. A plane or cut separating the body into left and right sections
6. Toward the head region of the body

Up:

3. The side surfaces of the body
7. Toward the foot region of the body
14. Opposite of bad

ANSWERS (These terms are to be used in filling in the crossword):

| dorsal | medial | right | frontal | superior | nothing | proximal | inferior | good |
| posterior | lateral | love | distal | anterior | transverse | sagittal | ventral | |

Name: _____   **Date:** _____

Diseases: Name That Category

Read each list of hints and determine the diseases to which they are referring. Write the name of the disease in the appropriate square. (*Note:* Answers are found in the list at the bottom of the woorksheet.)

1. Caused by the HIV virus; destroys the body's immune system
2. Commonly called lockjaw; caused by a bacterium; puncture wound
3. Body not producing enough insulin; high blood levels of glucose
4. Arteries narrow; plaques containing cholesterol build in arteries
5. Coronary artery in the heart becomes blocked; part of heart muscle is deprived of oxygen
6. Virus caused; fever; runny nose; dry cough; red spots develop on body
7. Virus caused; fever; sore throat; tired; weak; swollen lymph glands

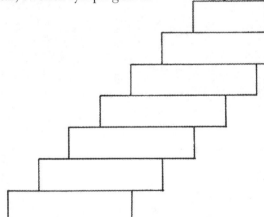

8. Parotid glands near ear swell; fever; diarrhea
9. Bacteria caused; runny nose; severe coughing
10. Rash on skin; reddish blisters; itching
11. Chills, fever; sore throat; headache
12. Caused by bacterial toxins; abdominal cramps; nausea and vomiting; diarrhea

Answers:

chicken pox	AIDS	mumps	food poisoning
tetanus	measles	atherosclerosis	influenza
whooping cough	diabetes	myocardial infarction	mononucleosis

WORKSHEET 5-40

Diseases: Name That Category

Read the hints and determine the proper category to which each belongs. Write the category in the appropriate box. Categories can be found in the list at the bottom of the worksheet.

1. Another name for skin cancer
2. Pouches form in walls of large intestine and occasionally become inflamed
3. An acute inflammation of the tonsils
4. Excessive fluid pressure build-up in the eye
5. Lungs lose elasticity and breathing becomes labored
6. Genetically caused high blood levels of cholesterol

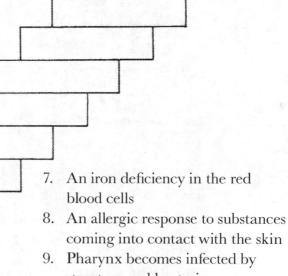

7. An iron deficiency in the red blood cells
8. An allergic response to substances coming into contact with the skin
9. Pharynx becomes infected by streptococcal bacteria
10. Clouding of lens of eye
11. Inflammation of stomach lining
12. Fungal infection of the skin

Categories:

emphysema	anemia	dermatitis	familial hypercholesterolemia
tonsilitis	strep throat	ringworm	diverticulosis
gastric ulcer	glaucoma	cataract	epithelial carcinoma

WORKSHEET 5-41

Terms Associated with Disease-Causing Organisms: Matching

a. antiseptic	f. pathogenic	k. sterile	p. vectors
b. immunity	g. incubation	l. antibiotic	q. toxin
c. communicable	h. virulence	m. chemotherapy	r. aseptic
d. stasis	i. pandemic	n. bacteriocide	s. fungi
e. sanitizing	j. disinfectant	o. epidemic	t. endemic

Select the terms listed above to match the statements that follow. Write the letter of the term in the blank.

_____ 1.   Free of all living organisms

_____ 2.   Technique used in working with microorganisms that assures that neither you nor your environment becomes contaminated

_____ 3.   Used to destroy disease-causing organisms on nonliving materials

_____ 4.   Used to destroy or prevent growth of disease-causing organisms on living materials

_____ 5.   Term describing an organism capable of causing disease

_____ 6.   Any agent that kills bacteria

_____ 7.   Suffix that means to prevent or retard the growth of a microscopic organism

_____ 8.   Produced by living organisms and used to fight disease organisms

_____ 9.   Substance produced by microorganisms that can cause sickness

_____ 10. The body's resistance to a specific disease

_____ 11. The treatment of disease using a chemical agent

_____ 12. Type of disease that can be transferred from one organism to another

_____ 13. Ability of an organism to cause disease

_____ 14. Worldwide epidemic of a disease

_____ 15. Reduces the number of organisms on an inanimate surface

_____ 16. Agents such as insects that can transfer disease-causing organisms from one organism to another

_____ 17. An unusual or sudden incidence of a disease

_____ 18. The period of time needed for a disease-causing organism to become established in the body

WORKSHEET 5-42

The Ear: Fill in the Blanks

Select the proper word or statement listed below to fill in the blanks of the sentences that follow.

incus	oval window	equalize	semicircular canals	vestibule
auditory nerve	Organ of Corti	cochlea	sound vibrations	malleus
tympanic membrane	auricle	tinnitus	four	cerumen
amplify and carry	Eustachian	ossicles	three	stapes

1. The ear is divided into _____ major sections.

2. The function of the ossicles is to _____ sound vibrations through the middle ear.

3. Another name for the eardrum is the _____.

4. The _____ in the inner ear houses the structure that converts sound vibrations into nerve impulses.

5. The collective name for the three tiny bones found in the middle ear is _____.

6. The name of the first of the three tiny bones found in the middle ear is _____.

7. The name of the second of the three tiny bones found in the middle ear is _____.

8. The name of the third of the three tiny bones found in the middle ear is _____.

9. _____ located in the inner ear deal with balance and equilibrium.

10. A tube connecting the back of the throat with the middle ear is the _____.

11. _____ is a term used to denote "ringing" in the ears.

12. A structure called the _____ carries nerve impulses from the inner ear to the brain.

13. The flap of the outer ear is called the _____.

14. The main task of the Eustachian tube is to _____ the air pressure in the middle ear to the air pressure of the outer ear.

15. The auricle's function is to direct _____ into the canal of the outer ear.

WORKSHEET 5-43

Endocrine Gland Hormones: Matching

Select the gland from the right-hand column that secretes the hormone in the left-hand column. Write the letter of the gland in the blank. Each gland might be used more than once.

_____ 1. Chorionic gonadotropin, estrogen, and progesterone (secreted only during pregnancy)

_____ 2. Thymosin

_____ 3. Glucocorticoids   a. adrenal

_____ 4. Insulin   b. gastric mucosa

_____ 5. Glucagon   c. pituitary

_____ 6. Gastric   d. ovaries

_____ 7. Somatotropin   e. testes

_____ 8. Calcitonin   f. thyroid

_____ 9. Parathyroid   g. parathyroid

_____ 10. Prolactin   h. islands of Langerhans

_____ 11. Melatonin   i. placenta

_____ 12. Luteinizing   j. pineal

_____ 13. Estrogen   k. thymus

_____ 14. Epinephrine

_____ 15. Thyroid

_____ 16. Testosterone

_____ 17. Oxytocin

_____ 18. Antidiuretic

_____ 19. Mineralocorticoids

_____ 20. Progesterone

WORKSHEET 5-44

Functions of Endocrine Gland Hormones: Matching

Matching: Select the hormone in the right-hand column that corresponds to the function given in the left-hand column. Write the letter of the term in the blank.

_____	1.	Development of the immune system	a. mineralocorticoids
_____	2.	Stimulates milk secretion by the mammary glands	b. glucagon
_____	3.	Induces ovulation	c. insulin
_____	4.	Promotes breast development during pregnancy	d. corionic gonadotropins
_____	5.	Promotes maleness	e. thymosin
_____	6.	Deals with development of the mammary gland during puberty	f. somatotropin
_____	7.	Regulates sugar intake by the cells of the body	g. calcitonin
_____	8.	Helps regulate the amount of calcium in the blood	h. prolactin
_____	9.	Regulates body growth and development (two answers)	i. parathyroid
_____	10.	Aids in increasing blood sugar if it falls too low	j. gastrin
_____	11.	Released during stress to prepare the body for action	k. glucocorticoids
_____	12.	Deals with fat catabolism by the cells	l. luteinizing
_____	13.	Deals with secretion of gastric juice in the stomach	m. epinephrine
_____	14.	Involved in regulating the female reproductive cycle	n. testosterone
_____	15.	Helps to regulate the blood level of calcium	o. estrogen
_____	16.	Stimulates milk ejection by the mammary glands	p. oxytocin
_____	17.	Aids in decreasing water loss from the body	q. antidiuretic
_____	18.	Regulates mineral metabolism	r. thyroid

WORKSHEET 5-45

Location of Endocrine Glands: Message Square

Cover the left column with a 3 × 5 card. Reveal one letter at a time, and write it in the proper square(s).

Hint: The general name for chemicals secreted by endocrine glands

Put: an N in 6.
        an E in 7.
        an R in 3.
        an S in 8.
        an O in 2 and 5.
        an M in 4.
        an H in 1.

1	2	3	4	5	6	7	8

RESEARCH QUESTIONS FOR TEXT OR LIBRARY

ANSWERS

SOURCE (INCLUDE PAGES)

1. Name the location for each of the following endocrine glands:
   - a. adrenal _____
   - b. gastric mucosa _____
   - c. pituitary _____
   - d. ovary _____
   - e. islands of Langerhans _____
   - f. thyroid _____
   - g. thymus _____
   - h. testes _____
   - i. placenta _____
   - j. parathyroid _____
   - k. pineal _____

2. What name is given to the organ upon which the endocrine gland has an effect?

Name: _____   **Date:** _____

Environmental Adaptation: Message Square

Cover the left column with a 3 × 5 card. Reveal one letter at a time, and write it in the proper square(s). Write the anticipated answer in the right column.

Hint: Two types of environmental adaptations by organisms

Anticipated Answer

Put: an O in 6, 8, and 20.
an L in 7, 13, and 23.
an H in 2 and 16.
an I in 5, 10, and 19.
an A in 12, 17, and 22.
an R in 21.
a P in 1.
a V in 18.
a Y in 3.
a B in 14.
a G in 9.
an S in 4.
an E in 15.
a C in 11.

RESEARCH QUESTIONS FOR TEXT OR LIBRARY

ANSWERS   SOURCE (INCLUDE PAGES)

1. What is mimicry?   _____

2. What is the value of mimicry?   _____

3. Give an example of mimicry. _____

4. Give an example of an animal behavioral adaptation. _____

5. Give an example of a plant behavioral adaptation. _____

WORKSHEET 5-47

Environmental Adaptation: Message Square

Cover the left column with a 3 × 5 card. Reveal one letter at a time, and write it in the proper square(s). Write the anticipated answer in the right-hand column.

Hint 1: Specialized cells that allow some animals to change color

1	2	3	4	5	6	7	8	9	10	11	12	13	14

Anticipated Answer

Put: an R in 3 and 12.
      an S in 14.
      an O in 4, 8, and 11.
      a T in 7.
      a C in 1.
      an E in 13.
      an H in 2 and 10.
      a P in 9.
      an M in 5.
      an A in 6.

Hint 2: A type of camouflage in which the organism blends in with its environment

1	2	3	4	5	6	7	■	8	9	10	11	12	13	14	15	16	17

Put: a C in 1, 7, and 8.
      an I in 6 and 15.
      an L in 10.
      an N in 17.
      an O in 9, 11, and 16.
      a Y in 3.
      an R in 2 and 12.
      a T in 5 and 14.
      an A in 13.
      a P in 4.

RESEARCH QUESTIONS FOR TEXT OR LIBRARY

ANSWERS   SOURCE
                 (INCLUDE PAGES)

1. What is warning coloration?   _____

2. What is the value of warning coloration?   _____

WORKSHEET 5-48

Environment and Pollution: Matching

Select the proper terms listed below to match the statements that follow. Write the letter of the term in the blank.

a. biomagnification	f. ecology	k. sulfur dioxide	o. flora
b. fauna	g. biodegradable	l. eutrophication	p. abiotic
c. thermal	h. ozone	m. smog	q. fossil
d. recycling	i. natural resources	n. biotic	r. pesticides
e. sulfuric	j. phosphates		

_____ 1. Process by which wastes are converted into new products and materials

_____ 2. The Earth's materials that are used by living things for such needs as food, shelter, and manufacturing

_____ 3. The toxic air-polluting gas that causes acid rain

_____ 4. The term used to denote increasing concentration of pesticides and other pollutants in the food chain

_____ 5. Term applied to wastes that break down into harmless substances when exposed to the environment

_____ 6. The layer of the atmosphere thought to be breaking down because of air pollution

_____ 7. A collective term for animals in the environment

_____ 8. The study of the interaction of living organisms with their environment

_____ 9. The nonliving factors of an organism's environment

_____ 10. Chemicals used to kill unwanted insects

_____ 11. Type of pollution in which water is heated

_____ 12. A collective term for plants in the environment

_____ 13. The acid in acid rain

_____ 14. The living factors of an organism's environment

_____ 15. Process which can lead to the death of a body of water from pollutants

_____ 16. An environmental polluting agent found in some detergents

_____ 17. Fog-like pollution caused by automobile emissions

_____ 18. Types of fuel, such as coal, that cause air pollution when burned

WORKSHEET 5-49

Enzymes: Word Scramble

Read each clue and then unscramble the words in parentheses, writing the answers in the blanks provided.

1. What enzymes are made of
   (RTEPION) __ __ __ __ __ __ __
                        4

2. What activities enzymes activate
   (HMCALCIE) __ __ __ __ __ __ __ __   (CTRAEOIN) __ __ __ __ __ __ __ __
                                                          7

3. The substance an enzyme acts upon
   (SBTSRUTAE) __ __ __ __ __ __ __ __ __

4. Class or group of enzymes that break down carbohydrates
   (MLAYAESS) __ __ __ __ __ __ __ __ __
                        1

5. Class or group of enzymes that break down proteins
   (RTESPAESO) __ __ __ __ __ __ __ __ __
                        6

6. Class or group of enzymes that break down fats
   (PLASEIS) __ __ __ __ __ __ __

7. Name of the hypothesis that helps to explain enzyme functioning
   (COLK) (NAD) __ __ __ __   __ __ __   (EYK) __ __ __

8. Vitamins act as these when they attach themselves to an enzyme.
   (ZMESYNECO) __ __ __ __ __ __ __ __ __

9. Enzymes lower this type of energy in a cellular reaction.
   (CIATATONIV) __ __ __ __ __ __ __ __ __ __ __
                        3

10. Enzymes act as these because they are not changed during a reaction.
    (TALSYSTAC) __ __ __ __ __ __ __ __ __
                                    9

11. Enzymes control this aspect of chemical reactions.
    (TARE) __ __ __ __
                8

12. The special region of an enzyme that joins with the substrate
    (CTAIEV) __ __ __ __ __ __   (IEST) __ __ __ __

13. The series of enzyme-controlled reactions that build complex material from simpler material
    (NABALOSIM) __ __ __ __ __ __ __ __ __ __

14. A change in the shape of an enzyme allowing it to react effectively with a substrate
    (DCDNUIE) __ __ __ __ __ __ __   (ITF) __ __ __
                    2   10   5

To learn the name of the enzyme that produces the glow of a firefly, select the appropriate numbers from above and write them on the dotted lines:

__ __ __ __ __ __ __ __ __ __
1 2 3   4 5 6 7 8   9 10

Name: _____   **Date:** _____

Experimentation: Message Square

Cover the left column with a 3 × 5 card. Reveal one letter at a time, and write it in the proper square(s). Write the anticipated answer in the right-hand column.

Hint: Scientists conduct experiments under these conditions.

1	2	3	4	5	6	7	8	9	10

Anticipated Answer

Put: an O in 2 and 6.
a D in 10.
an E in 9.
an N in 3.
an R in 5.
a C in 1.
a T in 4.
an L in 7 and 8.

RESEARCH QUESTIONS FOR TEXT OR LIBRARY   ANSWERS   SOURCE (INCLUDE PAGES)

1. Explain the answer to the above hint.   _____

2. What two groups are usually used in conducting an experiment?   _____

3. What is a hypothesis?   _____

4. What is a theory?   _____

5. What is data?   _____

6. What is the term used for studying data in determining what has been learned from an experiment?   _____

WORSHEET 5-51

Structure of the Eye: Labeling

Using the information presented at the bottom of this worksheet, label the numbered parts on the following diagram of the eye.

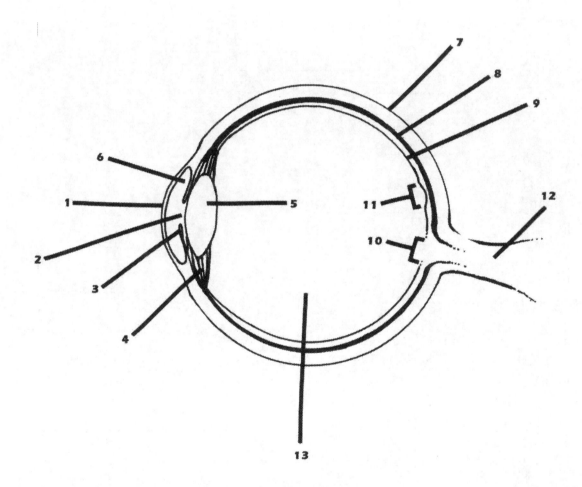

pupil	optic nerve	sclera layer	cornea
iris	choroid layer	vitreous gel (body)	aqueous humor
optic disc (blind spot)	lens	fovea	retina layer
suspensory ligaments			

Name: _____ **Date:** _____

Eye Structure and Function: Anagrams

The answer to each of the following statements is provided in anagram form to the left of each statement. Unscramble each anagram and write the answer just above it. (*Note:* The first two letters of each anagram word are the first two letters, in order, of the unscrambled answer.)

1. <u>PUIPL</u>   A hole through the iris that lets light enter the eye

2. <u>FOEAV</u>   A very small area on the retina of sharpest focus of incoming light rays

3. <u>BLNDI SPTO</u>   An area toward the back of the retina that does not register incoming light rays

4. <u>IRSI</u>   The colored part of the eye which regulates the amount of light entering the eye

5. <u>CHOORDI</u>   The middle layer of the eyeball which contains a black pigment and prevents extraneous light from "reflecting around" inside the eye

6. <u>VIRTOSU BOYD</u>   A gell-like material that fills the posterior section of the eyeball, giving it shape

7. <u>OPITC NEVRE</u>   Exiting at the back of the eye, this structure carries nerve impulses for vision to the brain.

8. <u>AQEUSOU HUOMR</u>   The liquid that fills the anterior section of the eyeball, giving it shape

9. <u>COJNUNCTVIA</u>   A thin mucous membrane layer that provides lubrication and covers the cornea and posterior surfaces of the eyelids

10. <u>SCRLEA</u>   The outer layer of the eyeball that provides protection

11. <u>CONREA</u>   The transparent part of the sclera layer, it is located on the front surface of the eyeball and aids in focusing light rays as they enter the eye.

12. <u>REITNA</u>   The inner layer of the eyeball that is made of light-sensitive nerve cells called rods and cones

13. <u>SUPENSSROY LIAGENTSM</u>   The structures that hold the lens in place

14. <u>LESN</u>   This structure aids in focusing light rays on the retina.

15. <u>ROSD</u>   Retinal nerve cells sensitive to low levels of light intensity but not to color

16. <u>COSEN</u>   Retinal nerve cells used for color vision

Name: _____   Date: _____

Fields of Biology: Anagrams

Use the definitions in the left-hand column to unscramble the anagrams in the middle column. Write the unscrambled words in the blanks provided in the right-hand column. (*Note:* The first three letters of each anagram are the first three letters, in order, of the unscrambled word.)

DEFINITIONS	ANAGRAMS	ANSWERS
1. The study of plants	botnya	__ __ __ __ __ __ 10 1
2. The study of animals	zooylg	__ __ __ __ __ __ __ 11 14 15
3. The study of microorganisms	micblroyoiog	__ __ __ __ __ __ __ __ __ __ __ __ 8 21 9
4. The study of the life and other aspects of the ocean	ocegnhypaoar	__ __ __ __ __ __ __ __ __ __ 19 3 2
5. The study of fungi	mycgoyol	__ __ __ __ __ __ __ __ 6
6. The study of insects	entglmoyoo	__ __ __ __ __ __ __ __ __ __ 4
7. The study of fish	ichlyhgyoto	__ __ __ __ __ __ __ __ __ __ __ 22
8. The study of the functioning of organisms	phyogyolis	__ __ __ __ __ __ __ __ __ __ 5 7
9. The study of the relationship between organisms and their environments	ecogoyl	__ __ __ __ __ __ __ 17 12 16
10. One who studies the role of diet health	nutointirsti	__ __ __ __ __ __ __ __ __ __ __ __ 18 20 13 23

To answer the following, select the appropriate numbered letters above and write them in the spaces to complete each statement.

a. The study of the structure of living things is __ __ __ __ __ __ __.

1 2 3 4 5 6 7

b. The study of the cell is __ __ __ __ __ __ __ __.

8 9 10 11 12 13 14 15

c. The study of the inheritance of genes is __ __ __ __ __ __ __ __.

16 17 18 19 20 21 22 23

WORKSHEET 5-54

Fish: True and False

Carefully read each statement. If the underlined word makes the statement true, write a T in the blank; if false, write an F. Correct each false statement by crossing out the word and writing the proper word, selected from the words listed at the bottom of the worksheet.

_____ 1. Fish are classified under phylum <u>Chordata</u>.

_____ 2. A type of biologist who studies animals in general is a <u>botanist</u>.

_____ 3. Fish are <u>invertebrates</u>.

_____ 4. Fish obtain their oxygen through structures called <u>gills</u>.

_____ 5. Fish are <u>warm</u>-blooded animals.

_____ 6. Most fish lay <u>hard-shelled</u> eggs.

_____ 7. Most fish lay eggs that are fertilized <u>externally</u>.

_____ 8. The fish heart has <u>four</u> chambers.

_____ 9. Bony fish belong to the class <u>Osteichthyes</u>.

_____ 10. Besides bony fish, two other groups are jawless and <u>cartilaginous</u>.

_____ 11. The <u>caudal</u> fin is located on the top of the body of the fish.

_____ 12. The protective covering of the gills of a fish is called the <u>spine</u>.

_____ 13. A live-bearing fish is termed <u>oviparous</u>.

_____ 14. <u>Fins</u> aid fish in locomotion.

_____ 15. A specialist in the study of fish is called an <u>ichthyologist</u>.

_____ 16. The <u>ventricle</u> is a structure found in bony fish that aids in buoyancy.

_____ 17. The <u>scale</u> of a fish detects sound vibrations and nearby movement.

_____ 18. The air-breathing <u>lungfish</u> is considered a "living fossil."

_____ 19. The <u>age</u> of a fish can be determined by counting the scale rings.

_____ 20. Whales, porpoises, and dolphins are all types of <u>fish</u>.

Words for correcting the false statements:

Porifera	entamologist	operculum	swim bladder	cold
dorsal	lungs	mycologist	ovoviparous	mammals
vertebrates	lateral line	soft-shelled	toadfish	lateral
soft	two	zoologist	internally	three

Name: _____   Date: _____

Anatomy of a Fish: Color Coding/Labeling

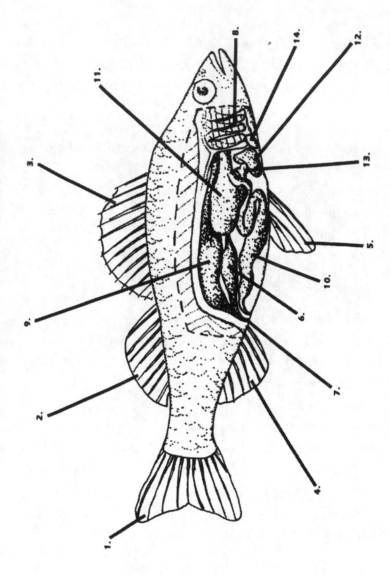

Key:

1. Caudal fin = red
2. Posterior dorsal fin = yellow
3. Dorsal fin = brown

4. Anal fin = blue
5. Pelvic fin = green
6. Reproduction organs = yellow

7. Anus = orange
8. Gills = leave white
9. Swim bladder = purple
10. Intestines = brown

11. Stomach = blue
12. Liver = orange
13. Gall bladder = purple
14. Heart = red

Name: _____ **Date:** _____

Structure of a Typical Flower:
Labeling/Fill in the Blanks

Use the terms below both to label the diagram of the flower and to fill in the blanks of the statements which follow.

stigma   anther   sepals   style   filament   ovary

petals   pistil (comprised of the stigma, style, and ovary)

stamen (comprised of the anther and filament)

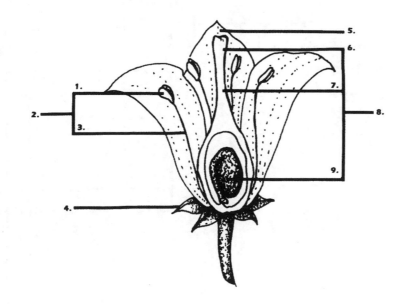

1. The _____ contains the egg cells.
2. The _____ is sticky and is located on top of the pistil.
3. The petals are surrounded by the _____.
4. The anther is supported by the _____.
5. The _____, a stalk-like structure, has the stigma on its tip and the ovary at its base.
6. Often brightly colored, the _____ are surrounded by the sepals.
7. Located at the top of the filament, the _____ contains the pollen.
8. The _____ is the collective term for the flower's female reproductive structures consisting of the stigma, style, and ovary.
9. The _____ is the collective term for the flower's male reproductive structures consisting of the anther and the filament.

Name: _____   Date: _____

The Food Chain: Crossword

Match the clues with the answers provided to complete the crossword.

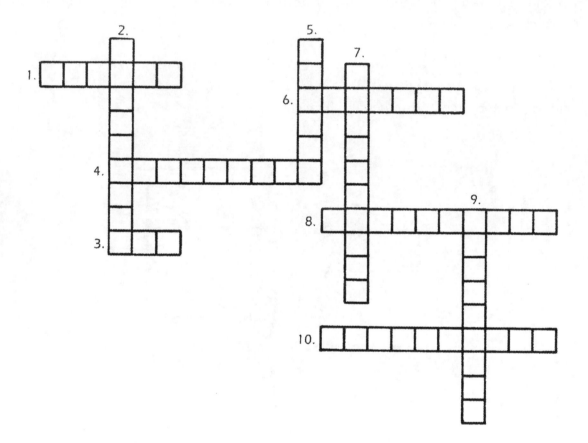

CLUES:

Across:
1. In a food chain, this is passed from organism to organism.
3. Original source of all energy in a food chain
4. Organisms that obtain food from other organisms
6. The type of organisms in a food chain that are called consumers
8. An animal that feeds on plants only
10. An animal that feeds on other animals only

Down:
2. Organisms in a food chain that can make their own food
5. Organisms in a food chain that are the producers
7. The organisms that feed directly off of plants are what part of the food chain? (two-word answer)
9. An animal that feeds on both plants and animals

ANSWERS:

herbivores	third level	sun	energy	bacteria	decomposers
producers	plants	omnivores	first level	consumers	carnivores
animals	sugars	air	carbon	nitrogen	fungi

Name: _____   **Date:** _____

Anatomy of a Frog: Color Coding/Labeling

Color code and label the parts of the frog diagrammed below using the key which follows the diagram.

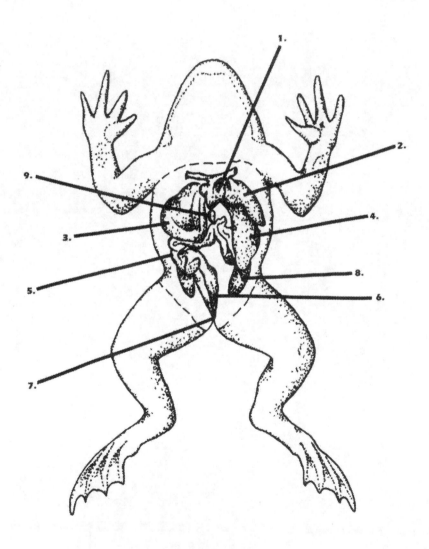

Key:

1. Heart = red
2. Lungs = blue
3. Liver = black

4. Stomach = purple
5. Small intestine = brown
6. Large intestine = green

7. Cloaca = blue
8. Kidney = red
9. Gall bladder = orange

WORKSHEET 5-59

Genetics: Matching

For each of the following groups of terms, connect each term in the left-hand column with its proper counterpart at the right by drawing a straight line between them.

Group 1:

a. heterozygous   a. One gene masking another's expression

b. multiple genes   b. Genetic makeup of an organism

c. multiple alleles   c. Having different alleles for a given trait

d. genotype   d. Two or more pairs of genes influencing a trait

e. phenotype   e. Physical characteristics of an organism

f. dominance   f. Three or more genes influencing a trait

Group 2:

a. crossing over   a. Genes on the same chromosome

b. nondisjunction   b. The exchange of genetic material between two chromosomes

c. polyploidy   c. Homozygous for a given trait

d. linkage   d. Failure of a chromosome pair to separate during meiosis

e. pure-bred   e. A cell having an extra set of chromosomes

f. blending inheritance   f. In a pair of genes, neither is dominant or recessive.

Group 3:

a. Mendel   a. Different forms of a particular gene

b. mutation   b. Used to diagram genetic crosses

c. homozygous   c. Fruit flies used in genetic research

d. alleles   d. Father of genetics

e. Drosophila   e. An error in the copying of a gene

f. Punnett square   f. Having identical alleles for a trait

WORKSHEET 5-60

Genetics: Sentence Correction

Each sentence has a boldfaced term that does not belong in that sentence but does belong in one of the other sentences. Correct each sentence by writing in the proper boldfaced term.

1. The study of the inheritance of traits is called **meiosis.**

2. When one gene expresses itself over another, the former gene is said to be the **DNA** gene.

3. Your behavior and physical appearance is referred to as your **mutation.**

4. The genes that you are carrying in your cells are referred to as your **dominant.**

5. The process by which a chromosome makes an identical copy of itself is called **heredity.**

6. When one gene expresses itself over another gene, the latter gene is said to be the **sex-linked trait** gene.

7. The process of cell division resulting in daughter cells, each of which contains one half of the normal number of chromosomes, is called **replication.**

8. Chromosomes are made of a material called **phenotype.**

9. When a chromosome does not make an accurate copy of itself, the result is called a **genotype.**

10. When a sperm cell (which carries chromosomes from the father) fertilizes an egg cell (which carries chromosomes from the mother), the resulting cell is called the **albino.**

11. The X and Y chromosomes are called the **genes.**

12. Chromosomes are comprised of hundreds of **sex chromosomes.**

13. The normal number of chromosomes in a human body cell is **23**.

14. A trait or disorder that is inherited through the sex chromosomes is called a **color blindness.**

15. An example of a sex-linked disorder is **recessive.**

16. The number of chromosomes carried in human sperm and egg cells is **46.**

17. The organism that, as a result of a mutation, cannot produce pigment for skin, hair, or eyes is called a **zygote.**

WORKSHEET 5-61

Genetic Diseases and Diagnosis: Word Scramble

Read each clue and then unscramble the words in parentheses, writing the answers in the blanks provided.

1. A sex-linked disease, it is the most common type of color blindness.
   (EDR) (REGEN) __ __ __   __ __ __ __ __
                     14   15

2. In this disease, the blood does not clot properly.
   (MEHOHIPLAI) __ __ __ __ __ __ __ __ __ __
                          8

3. A disease afflicting males in which normal sexual characteristics fail to develop
   (FTKINELERELS) __ __ __ __ __ __ __ __ __ __ __ __
                                    1
      (YDMSNROE) __ __ __ __ __ __ __ __

4. A disease that involves mental retardation and abnormal facial characteristics
   (WODNS) (SDERNYOM) __ __ __ __ __   __ __ __ __ __ __ __ __ __
                                                        10

5. A disease afflicting females in which normal female characteristics do not develop
   (SUNERTR) (MOESNRDY) __ __ __ __ __ __ __   __ __ __ __ __ __ __ __

6. A disease involving fat build-up in the brain cells
   (AYT) (ACSHS) __ __ __ __ __ __ __ __
                  12   6

7. A disease in which the red blood cells are misshapen and unable to carry oxygen effectively
   (IKCLES) (ECLL) __ __ __ __ __ __ __   __ __ __ __   (NMEIA) __ __ __ __ __ __

8. A disease in which the body is unable to metabolize a particular amino acid, resulting in mental retardation
   (EYKIAPNLHTEOUNR) __ __ __ __ __ __ __ __ __ __ __ __ __ __ __
                                  9

9. A genetic disease can be caused when this happens to a gene.
   (UATMTONI) __ __ __ __ __ __ __ __
                 3   7

10. A diagnostic technique which involves the insertion of a needle to obtain fetal cells
    (MONTSANISECEI) __ __ __ __ __ __ __ __ __ __ __ __ __
                     13 16   2   5

11. A diagnostic technique in which sound waves are used to obtain an image of a fetus
    (TRULDNSAUO) __ __ __ __ __ __ __ __ __ __
                  11   4

To answer the following, select the appropriate numbered letters from above and write them on the dotted lines to complete each statement.

a. The technique that allows one to see the fetus by using an endoscope is called
$\overline{}_1 \ \overline{}_2 \ \overline{}_3 \ \overline{}_4 \ \overline{}_5 \ \overline{}_6 \ \overline{}_7 \ \overline{}_8 \ \overline{}_9$.

b. The term for something in the environment capable of causing a gene mutation is
$\overline{}_{10} \ \overline{}_{11} \ \overline{}_{12} \ \overline{}_{13} \ \overline{}_{14} \ \overline{}_{15} \ \overline{}_{16}$.

WORKSHEET 5-62

The Human Heart: Around the Square

Using the clues and answers provided, fill in the squares with the appropriate answers beginning with number 1 at the lower left and proceeding clockwise. (Selected numbers around the square are provided as a guide.)

Clues:

1. The human heart has four of these.
2. The vein that drains blood into the right side of the heart
3. What the two lower chambers of the heart are called
4. What the two upper chambers of the heart are called
5. The name of the valve that separates the left atrium from the left ventricle
6. The name of the valve that separates the right atrium from the right ventricle
7. The name of the artery that carries blood from the left ventricle to the body (with the exception of the lungs)
8. The name of the artery that carries blood from the right ventricle to the lungs
9. The base of the heart
10. The term used for oxygen-rich blood
11. The name of the vein that carries blood from the lungs to the left atrium
12. The type of muscle tissue found only in the heart
13. One risk factor for heart disease is whether or not you have this habit.
14. Another name for a heart attack

Answers:

tricuspid	apex	ventricles	oxygenated
mitral	smoking	Aorta	cardiac
coronary	chambers	Vena Cava	atria
pulmonary (used twice)			

Immune System: Word Scramble

Read each clue and then unscramble the words in parentheses, writing the answers in the blanks provided.

1. The kind of blood cells that play a role in the body's immune system
    (TIEWH) __ __ __ __ __
                    15 6

2. A substance foreign to the body, such as a disease-causing organism, that stimulates the body's immune system
    (TNAGIEN) __ __ __ __ __ __ __
               1

3. A protein substance produced by the body to fight an invading foreign substance such as a disease-causing organism
    (NIOYBADT) __ __ __ __ __ __ __ __
                8

4. Immune system T-cells develop in this gland.
    (MHSYUT) __ __ __ __ __ __
                    18 14

5. A chemical produced by the body that protects against viruses
    (TRINRONFEE) __ __ __ __ __ __ __ __ __ __ __
                  17

6. The immune system can cause this response following an organ transplant.
    (ECNREOJIT) __ __ __ __ __ __ __ __ __
                 20   3   19

7. A chemical that can suppress the immune system following an organ transplant
    (LSRCNIYCOEOP) __ __ __ __ __ __ __ __ __ __ __ __
                                            11   12

8. A specific white blood cell that is involved in antibody production
    (MPHCYETOYL) __ __ __ __ __ __ __ __ __ __

9. The process by which a white blood cell ingests a disease-causing organism
    (HGTSOCPOYISA) __ __ __ __ __ __ __ __ __ __ __ __
                    7   13   16   10

10. The disease characterized by a complete breakdown of the body's immune system
    (SDAI) __ __ __ __
                4

11. A substance injected into the body that helps protect against disease
    (CIENCVA) __ __ __ __ __ __ __
               5   2

To answer the following, select the appropriate numbered letters from above and write them on the dotted lines to complete each statement.

a. The type of immunization that is long lasting is called __ __ __ __ __ __.
                                                            1   2   3   4   5   6

b. The type of immunity that is short term is __ __ __ __ __ V __.
                                         7   8   9 10 11   12

c. The type of disease in which the immune system itself attacks the body is called
__ __ __ __ __ __ M U __ __.
13 14 15   16 17 18   19 20

Name: _____   Date: _____

Laboratory Equipment: Crossword

Match the clues to the answers provided to complete the crossword.

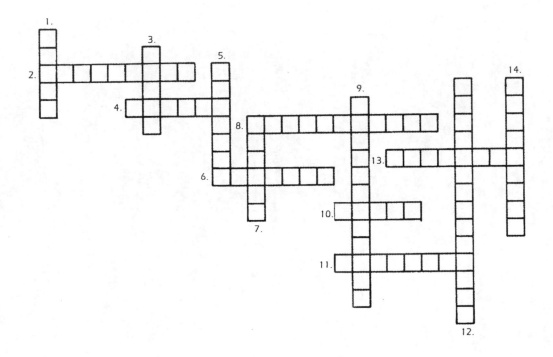

CLUES:

Across:
2. A container or enclosure used in growing plant ecosystems
4. A deep, wide-mouthed vessel for measuring liquids
6. Used for cutting plant or animal tissue
8. Used to listen to heart sounds
10. Model used for studying human internal organs
11. Used for storing, analyzing, and retrieving data
13. Type of automatic stirrer

Down:
1. Name of dish used for culturing bacteria
3. Used for measuring and transferring small amounts of liquid

5. Tweezer-like instrument for grasping
9. Instrument used for viewing electrocardiogram activity of the heart
14. Instrument used for sterilizing

Up:
7. Used to make cells and their structures more visible under the microscope
12. Used for projecting microscope slides onto a screen

ANSWERS (These terms are to be used in filling in the crossword):

magnetic	scalpel	computer	terrarium	oscilloscope
Petri	autoclave	torso	microprojector	forceps
Stethoscope	stains	pipet	beaker	

WORKSHEET 5-65

Laboratory Equipment: Message Square

Cover the left column with a 3 ×5 card. Reveal one letter at a time, and write it in the proper square(s). Write the anticipated answer in the right-hand column.

Hint 1: Two very important pieces of safety equipment for personal protection in the lab

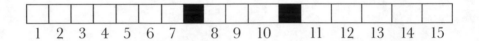

Anticipated Answer

Put: an S in 7.
       an N in 9 and 15.
       an R in 13.
       an O in 2 and 14.
       an E in 6.
       an L in 5.
       an A in 8 and 11.
       a G in 1, 3, and 4.
       a P in 12.
       a D in 10.

Hint 2: An instrument that could be used to observe light absorption characteristics of chlorophyll

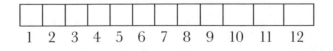

Put: a P in 2 and 11.
       an E in 3 and 12.
       an O in 7 and 10.
       an S in 1 and 8.
       a T in 5.
       a C in 4 and 9.
       an R in 6.

RESEARCH QUESTIONS FOR TEXT OR LIBRARY

ANSWERS

SOURCE
(INCLUDE PAGES)

1. Basically, how does the instrument in Hint 2 work? _____

2. What can be learned about stars using the instrument in Hint 2? _____

WORKSHEET 5-66

Laboratory Equipment: Message Square

Cover the left column with a 3 × 5 card. Reveal one letter at a time, and write it in the proper square(s). Write the anticipated answer in the right column.

Hint: Two popular types of microscopes used in the lab

Anticipated Answer

Put: an A in 8 and 18.
      an I in 12.
      a T in 2 and 7.
      an N in 6, 9, and 13.
      an S in 1.
      a B in 11.
      a D in 4 and 10.
      a C in 15.
      a U in 3 and 16.
      an L in 17.
      an E in 5.
      an O in 14.
      an R in 19.

RESEARCH QUESTIONS FOR TEXT OR LIBRARY	ANSWERS	SOURCE (INCLUDE PAGES)
1. What is the basic difference between the two types of microscopes in the hint above?		_____

2. What type of an electron microscope is able to provide three-dimensional type images?

_____

3. What does the electron microscope use for focusing the electron beam?

_____

WORKSHEET 5-67

Internal Structures
and Functions of a Leaf: Matching

Match the proper statement beginnings below to the statement endings which follow them. Write the proper letters in the blanks.

Statement Beginnings:

_____ 1.   The specialized tissue called phloem is used

_____ 2.   The specialized tissue called xylem is used

_____ 3.   The upper epidermis is

_____ 4.   Inside certain cells of the leaf are

_____ 5.   The waxy coating that functions to prevent excess water loss from the leaf is found on

_____ 6.   The lower epidermis is

_____ 7.   A layer of cells in which most photosynthesis takes place and located just below the upper epidermis is called

_____ 8.   Guard cells are structures

_____ 9.   The spongy layer is

_____ 10. Openings through the lower epidermis of the leaf, which allow gases such as oxygen, carbon dioxide, and water vapor to pass through,

_____ 11. The veins of the leaf contain the

_____ 12. In regulating the opening and the closing of the stomata, the guard cells change shape depending on

Statement Endings:

a.   is a layer of cells that covers the top surface of the leaf.
b.   to carry water and minerals from the roots to the leaves.
c.   xylem and the phloem.
d.   the palisade layer.
e.   a layer of cells that covers the bottom surface of the leaf.
f.   a layer of photosynthetic cells located beneath the palisade layer.

g. to carry food from the leaf to other parts of the plant.
h. the upper and lower epidermal layers of the leaf and is called cutin.
i. are called stomata.
j. structures called chloroplasts which carry out photosynthesis.
k. that regulate the opening and the closing of the stomata.
l. the amount of water that they gain or lose.

WORKSHEET 5-68

Leaf Structure: Labeling

Part 1: Internal Leaf Structure—Label the following diagram using the terms given below.

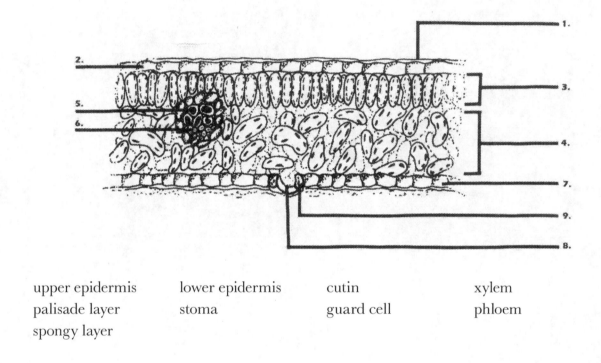

upper epidermis	lower epidermis	cutin	xylem
palisade layer	stoma	guard cell	phloem
spongy layer			

Part 2: External Leaf Structure—Label the following composite diagram using the terms given below.

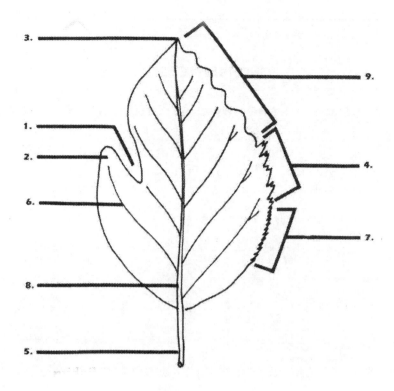

sinus   midrib   leaf stalk   wavy-edged

lobe   tip   vein   double-toothed

single-toothed

WORKSHEET 5-69

Deciduous Leaf Types and Arrangements: Matching

Match the proper terms given at the bottom of this worksheet with the diagrams that follow.
Write the terms in the blanks provided.

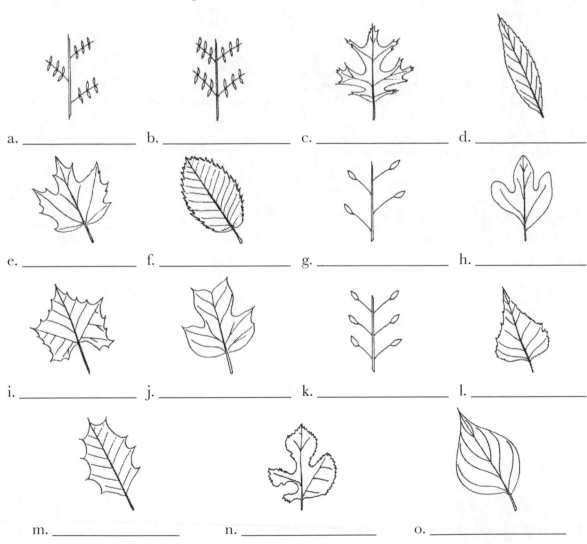

a. _____   b. _____   c. _____   d. _____

e. _____   f. _____   g. _____   h. _____

i. _____   j. _____   k. _____   l. _____

m. _____   n. _____   o. _____

alternate simple	alternate compound	opposite simple	opposite compound
Scarlet Oak	Flowering Dogwood	Sugar Maple	Sycamore
Tulip Tree	Sassafras	Mulberry	American Elm
Gray Birch	Weeping Willow	Holly	

WORKSHEET 5-70

Medical Science Terms: Anagrams

The answer to each of the following statements is provided in anagram form to the left of each statement. Unscramble each anagram and write the answer above it. (*Note:* The first three letters of each anagram are the first three letters, in order, of the unscrambled answer.)

1. DIANSOSIG: A determination of the type of disease afflicting an individual

2. INFCETOIN: A disease-causing organism growing within a host

3. HYPDOREIMC: Type of injection beneath or under the skin

4. PROIGNSOS: The forecast of the outcome of a disease or injury

5. IMMTYUIN: Protection against disease

6. CAROICNEGN: A cancer-causing agent

7. BIOSPY: Removal of tissue from the living body for examination

8. BENGNI: Not malignant

9. THRMBOSU: A blood clot

10. HYPTNONREIES: A term for high blood pressure

11. INSNMAIO: Difficulty sleeping

12. SYNODMRE: A disease characterized by several different symptoms

13. INCBUAOTNI: The time between becoming infected and showing symptoms

14. VIRSU: Organism that causes the common cold

15. STELEIR: Not contaminated by any known living organism

16. AMBAULOYRT: Able to move about

17. CORYNAOR: Relating to the heart

18. ORTPCSOHEDI: Relating to the bones and joints

19. DERTOOLYGAM: Relating to the skin

20. EPIMDEIC: An unusual and/or wide outbreak of a disease

21. STRKEO: A blood clot in the brain

22. GASTRIC: Relating to the stomach

WORKSHEET 5-71

Meiosis in Humans: Message Square

Cover the left-hand column with a 3 × 5 card. Reveal one letter at a time, and write it in the proper square(s). Write the anticipated answer in the right column.

Hint: The end products of meiosis

Anticipated Answer

Put: an A in 2, 5, and 11.
      an E in 4, 9, 13, 15, and 18.
      an M in 1 and 10.
      an S in 14 and 21.
      an X in 16.
      an L in 3, 12, 19, and 20.
      an N in 6.
      a C in 17.
      an F in 8.
      a D in 7.

RESEARCH QUESTIONS FOR TEXT OR LIBRARY	ANSWERS	SOURCE (INCLUDE PAGES)
1. What is the collective term used for the answer above?		_____
2. Why is meiosis referred to as reductive division?		_____

3. a. How many chromosomes are in the parent cell undergoing meiosis?

   _____

   b How many chromosomes will each daughter cell have at the completion of meiosis?

   _____

4. What is the name given to the female gametes?

   _____

5. What is the name given to the male gametes?

   _____

6. When a male gamete fertilizes a female gamete, how many chromosomes will the fertilized egg contain?

   _____

Name: _____   Date: _____

Mitosis: Matching

Match the questions below with the proper answers that follow them. Write the proper letter in the blanks.

Questions:

_____ 1. What is mitosis?

_____ 2. In order, what are the four main stages of mitosis?

_____ 3. What is the name of the stage a cell goes through just prior to mitosis?

_____ 4. What is the main event of interphase?

_____ 5. What are two important events of prophase?

_____ 6. What is the main event of metaphase?

_____ 7. What structure is involved in moving chromosomes during mitosis?

_____ 8. What is the main event of anaphase?

_____ 9. What are two important events of telophase?

_____ 10. At the completion of mitosis when the cell divides, what name is given to the two new cells?

_____ 11. You began life as a one-celled structure called a zygote. What process then took place over and over to build a body containing billions of cells?

Answers:

a. Interphase

b. The chromosomes in the nucleus of the cell make identical copies of themselves.

c. The chromosomes move toward and line up along the center of the cell, called the equator.

d. The nuclear membrane disappears and the chromosomes become distinct.

e. Mitosis

f. New nuclear membranes form around each of the two sets of chromosomes, and the cell divides into two daughter cells.

g. Prophase, metaphase, anaphase, and telophase

h. Daughter cells

i. Microtubules

j. Division of the nucleus of the cell (usually followed by division of the cell itself)

k. The microtubules pull one set of chromosomes to one side of the cell and an identical set to the opposite side of the cell.

Name: _____ _____ Date: _____

Muscle Groups and Actions: Multiple Choice

Select and circle the correct term in each of the following sentences.

1. The muscle contracted to flex your forearm is the (a. biceps, b. triceps, c. brachialis).

2. The (a. deltoid, b. triceps, c. pectoralis major) is contracted to abduct or lift your arms straight out to the sides.

3. To rotate or turn your head left and right, the (a. trapezius, b. sternocleidomastoid, c. pectoralis minor) is contracted.

4. To tilt your head back as in looking skyward, you must contract your (a. latissimus dorsi, b. deltoid, c. trapezius).

5. When the (a. platysma, b. corrugator, c. obicularis oculi) muscles are contracted, your eyes close.

6. The (a. diaphragm, b. transversalis, c. soleus) contracts as a part of the breathing process.

7. When doing pushups, the (a. triceps, b. biceps, c. serratus anterior) contracts, pushing your upper body away from the floor.

8. The hamstring group of muscles forms the flesh on the (a. anterior, b. posterior) surface of the thigh.

9. Which one of the following muscles does not belong to the hamstring group: (a. vastus lateralis, b. semitendinosus, c. biceps femoris)?

10. The (a. pectoralis minor, b. latissimus dorsi, c. teres minor) contracts to pull your shoulders forward as well as down.

11. To tilt your head forward as in looking down requires that you contract your (a. tibialis anterior, b. pectoralis major, c. sternocleidomastoid) muscle.

12. A contraction of your (a. obicularis oris, b. epicranius, c. masseter) closes your jaw.

13. The group of muscles known as the quadriceps or quads forms the flesh of the (a. anterior, b. posterior) surface of the thigh.

14. Which one of the following muscles does not belong to the quadriceps group: (a. sartorius, b. rectus femoris, c. vastus medialis)?

15. To move your legs out to the sides as in doing jumping jacks requires that your (a. adductor brevis, b. vastus lateralis, c. gluteus medius) contracts.

16. To bend your upper body forward, you must contract your (a. adductor magnus, b. iliopsoas, c. gluteus minimus) muscle.

Name: _____   **Date:** _____

Muscles of the Body: Fill in the Blanks

As you read the 21 hints, select the correct answers from the terms at the bottom of this worksheet and write them in the proper blanks.

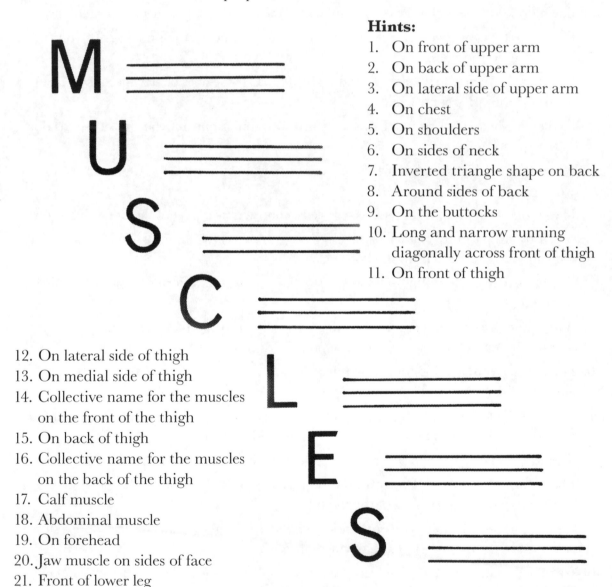

Hints:
1. On front of upper arm
2. On back of upper arm
3. On lateral side of upper arm
4. On chest
5. On shoulders
6. On sides of neck
7. Inverted triangle shape on back
8. Around sides of back
9. On the buttocks
10. Long and narrow running diagonally across front of thigh
11. On front of thigh
12. On lateral side of thigh
13. On medial side of thigh
14. Collective name for the muscles on the front of the thigh
15. On back of thigh
16. Collective name for the muscles on the back of the thigh
17. Calf muscle
18. Abdominal muscle
19. On forehead
20. Jaw muscle on sides of face
21. Front of lower leg

Terms:

sartorius	vastus lateralis	biceps femoris	hamstrings
tibialis	sternocleidomastoid	deltoid	masseter
brachialis	latissimus dorsi	rectus abdominus	triceps
frontalis	gastrocnemius	pectoralis major	quadriceps
rectus femoristrapezius	vastus medialis	biceps	
gluteus maximus	palmaris longus	diaphragm	platysma

WORKSHEET 5-75

The Nervous System: True and False

Read each statement carefully. If the boldfaced word(s) makes the statement true, write a T in the blank; if false, write an F. Correct the false statements by crossing out the boldfaced word(s) and writing the correct word above it.

_____ 1.   **Sensory** neurons carry information from skin receptors to the brain and spinal cord.

_____ 2.   The brain and spinal cord constitute the **peripheral** nervous system.

_____ 3.   The three principal parts of a neuron are a dendrite, an axon, and a **nerve impulse**.

_____ 4.   The minute gap between two neurons is called a **synapse.**

_____ 5.   The brain is wrapped in three layers of tissue known as the **neurolemma.**

_____ 6.   The cerebrum of the brain has **six** visible lobes.

_____ 7.   There are 12 pairs of **spinal** nerves attached to the brain.

_____ 8.   An involuntary action of the nervous system is called a **reflex.**

_____ 9.   Motor neurons carry information **away from** muscles and glands.

_____ 10. Wrapped around the axon of a neuron, the **myeline sheath** serves the function of insulation.

_____ 11. Another name for a nerve impulse is an **action potential.**

_____ 12. A lumbar puncture procedure is used to obtain a sample of **nerve tissue.**

_____ 13. The outer cortex of the cerebrum is composed of **white** matter.

_____ 14. The **corpus callosum** connects the right and left hemispheres of the brain.

_____ 15. A chemical released at the synapse is called a **neurotransmitter.**

_____ 16. The olfactory nerve deals with the sense of **hearing.**

_____ 17. A record of brain activity is the **electroencephalogram.**

_____ 18. The nervous system basically acts as a **communication** system.

_____ 19. The spinal cord is protected by bones called **modules.**

_____ 20. A multipolar neuron has two or more **axons.**

WORKSHEET 5-76

Nutrition: Fill in the Blanks

Select the proper word listed below to fill in the blanks of the sentences that follow.

exercise	vitamin	nine	generic	gain	150 mg
fiber	sugar	fat	liver	greatest	proteins
saturated	calories	four	nutritionist	200 mg	six

1. Each gram of fat in the diet provides _____ calories.

2. Of the three main types of fat in the diet, _____ are considered the least desirable.

3. All of the cholesterol we need on a daily basis is produced by an organ of the body called the _____.

4. Each gram of carbohydrate or protein in the diet provides _____ calories.

5. To lose weight more effectively one should decrease the amount of _____ in one's diet.

6. _____ is an excellent activity in addition to a prudent diet for weight loss.

7. Many medical experts recommend that a teenager's serum cholesterol level be _____ or lower.

8. A substance required by the body in small quantities for metabolism is called a _____.

9. Food products sold with no brand name on the label are referred to as _____ products.

10. On product labels, the ingredients are listed in order of quantity with the first ingredient being in _____ quantity.

11. The amount of energy provided by a food is measured in units called _____.

12. If you take in, over a period of time, more calories per day than you use for metabolism then you will _____ weight.

13. A person who specializes in diet and its effects on the body is a _____.

14. Amino acids are needed by the body to build _____.

15. Many nutritionists feel that in addition to fat, on a daily basis we take in too much salt and _____.

16. One of the roles of _____ in the diet is to aid in moving food along the digestive tract.

Name: _____   Date: _____

Phobias: Name That Category

Read the hints and determine the proper category to which each belongs. Write the category in the appropriate box. Categories can be found in the list at the bottom of the worksheet.

1. Fear of confined spaces
2. Fear of water
3. Fear of strangers
4. Fear of spiders
5. Fear of heights
6. Fear of darkness

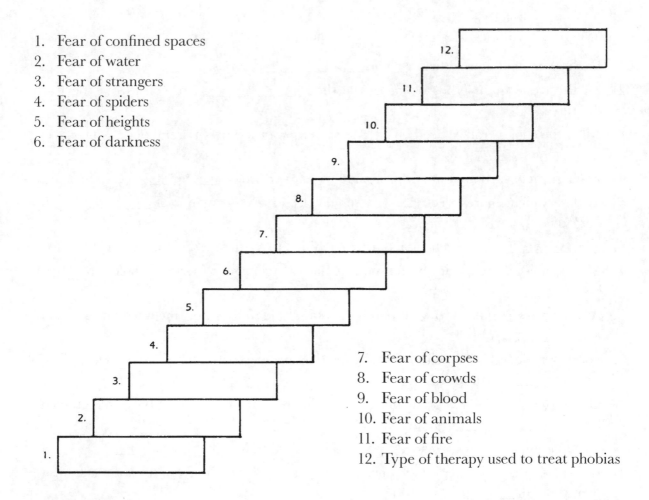

7. Fear of corpses
8. Fear of crowds
9. Fear of blood
10. Fear of animals
11. Fear of fire
12. Type of therapy used to treat phobias

Categories:

pyrophobia	agoraphobia	nyctophobia	claustrophobia
zoophobia	hydrophobia	xenophobia	arachnophobia
necrophobia	acrophobia	hemophobia	behavioral modification

WORSHEET 5-78

Photosynthesis: Around the Square

Using the clues and answers provided, fill in the squares with the appropriate answers beginning with number 1 at the lower left and proceeding clockwise. (Selected numbers around the square are provided as a guide.)

Clues:

1. Cellular _____ could not occur without photosynthesis.
2. This substance can trap light energy and convert it to chemical energy.
3. The two categories of reactions that occur during photosynthesis (two terms in answer)
4. The two main reactants used in photosynthesis (two terms in answer)
5. The two main products of photosynthesis (two terms in answer)
6. Chloroplasts have several of these membrane structures arranged in stacks.
7. In leaves, the spongy layer, along with this layer, carries out photosynthesis.
8. Chlorophyll absorbs heavily these two colors in the spectrum, (two terms in answer)
9. The form in which plants store the sugar produced by photosynthesis
10. This category of organisms carries out photosynthesis.

Terms:

chlorophyll	starch	respiration	producers	light
oxygen	green	grana	dark	water
glucose	palisade	violet	red	bacteria
carbon dioxide	yellow	protozoans	crista	sucrose

WORKSHEET 5-79

Plant Growth Responses: Word Scramble

Read each clue and then unscramble the words in parentheses, writing the answers in the blanks provided.

1. Any environmental factor that elicits a response in a plant
   (TMUSUSIL) __ __ __ __ __ __ __ __
              3   21   15

2. If the opposite sides of a plant stem are exposed to a stimulation in unequal amounts, this type of response is elicited.
   (PMTSROI) __ __ __ __ __ __ __
             17 7

3. Two groups of plant growth hormones
   (UINXAS __ __ __ __ __ __   (BINGLEIERSBL) __ __ __ __ __ __ __ __ __ __ __
           2   10   5   18

4. Term for a response to gravity
   (TRMSGEPOIO) __ __ __ __ __ __ __ __ __

5. Term for a response to light
   (TMPHSITOPORO) __ __ __ __ __ __ __ __ __ __ __ __
                  12

6. Term for the response of a flowering plant to periods of dark and light
   (HTPMPDIOSIOEOR) __ __ __ __ __ __ __ __ __ __ __ __ __ __
                                                      24

7. Term for a response to chemicals
   (HMISPMCETROO) __ __ __ __ __ __ __ __ __ __ __ __
                  6   23

8. Term for a response to touch
   (HMOMTOPISIRGT) __ __ __ __ __ __ __ __ __ __ __ __ __
                   4 16   11   22

9. Exposing a plant to cold to cause flowering
   (NOVNEIARLZTAI) __ __ __ __ __ __ __ __ __ __ __ __
                   1   8

10. Slowing down of growth when a plant is exposed to unfavorable environmental conditions
    (OMDAYCRN) __ __ __ __ __ __ __
                        9   14

11. A leaf pigment sensitive to light
    (HOREOHPYMCT) __ __ __ __ __ __ __ __ __ __ __
                           19   13

12. Name given to a hypothetical hormone thought to be responsible for flowering
    (LFNGIEOGR) __ __ __ __ __ __ __ __ __
                           20

To answer the following, select the appropriate numbered letters from above and write them on the dotted lines to complete each statement.

a. Plant movements that do not depend on the direction that a stimulus is coming from are called

$\overline{}$ $\overline{}$ $\overline{}$ $\overline{}$ $\overline{}$ $\overline{}$
 1   2   3   4   5   6

b. Biology specialists who might study plant responses to environmental stimuli are called

$\overline{}$ $\overline{}$ $\overline{}$ $\overline{}$ $\overline{}$ $\overline{}$ $\overline{}$ $\overline{}$ $\overline{}$ $\overline{}$ $\overline{}$ $\overline{}$ $\overline{}$ $\overline{}$ $\overline{}$ $\overline{}$ $\overline{}$
 7   8   9   10 11   12 13 14 15 16 17 18 19 20 21 22 23 24

WORKSHEET 5-80

The Higher Plants: Message Square

Cover the left column with a 3 × 5 card. Reveal one letter at a time, and write it in the proper square(s). Write the anticipated answer in the right column.

Hint 1: A collective term for plants that have specialized structures for transporting food and water throughout the plant

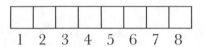

Anticipated Answer

Put: an R in 8.
        an A in 3 and 7.
        a C in 4.
        a V in 1.
        an L in 6.
        an S in 3.
        a U in 5.

Hint 2: Two kinds of specialized structures used for transporting water and food throughout the plant

Anticipated Answer

Put: an E in 4 and 10.
        an L in 3 and 8.
        an M in 5 and 11.
        an O in 9.
        an X in 1.
        an H in 7.
        a P in 6.
        a Y in 2.

RESEARCH QUESTIONS FOR TEXT OR LIBRARY

ANSWERS

SOURCE
(INCLUDE PAGES)

1. What is the name of the phylum for the type of plant in Hint 1?

_____

2. Most of the plants that you are familiar with are of the type in Hint 1. What are the three main groups of these types of plants?

_____

3. What is the function of each of the two specialized structures in Hint 2?

_____

Name: _____

Date: _____

WORKSHEET 5-81

Protozoans: Crossword

Instructions: Match the clues and answers provided to complete the crossword.

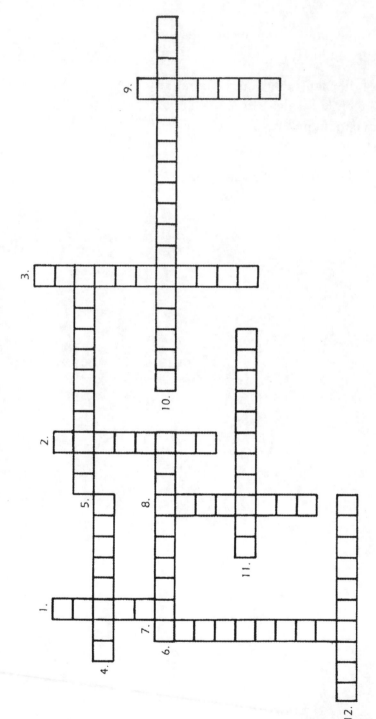

CLUES:

Across:

4. The classification kingdom to which protozoans belong
5. Protozoans are members of this important collection of microorganisms.
6. Protozoan referred to as the "slipper organism."
10. A structure (two words) found inside many protozoans that serves the function of pumping excess water out of the organism.
11. Common name for that group of protozoans that use flagella for locomotion
12. The genus name for specific protozoans that can cause malaria

Down:

1. A specific protozoan that is capable of constantly changing its shape as it moves and explores its environment
2. The term used to describe the food chain role of protozoans
3. One technique by which a paramecium reproduces
7. Means false feet
8. Common name for that group of protozoans that use cilia for locomotion
9. Name of structure that regulates cell activities.

ANSWERS:

Protista	zooplankton	amoeba	consumer	ciliates	paramecium
flagellates	Plasmodium	pseudopods	conjugation	nucleus	contractile vacuole

WORKSHEET 5-82

The Respiratory System: Multiple Choice

Circle the correct term in each of the following sentences.

1. The portal of entry and exit for air that is a part of the respiratory system is (a. the mouth, b. the nose, c. a combination of the mouth and nose).

2. Tiny air sacs in the lungs that are involved in the exchange of gases are called (a. bronchial sacs, b. alveolar sacs, c. pleural sacs).

3. Another name for the "wind pipe" is the (a. alveolar duct, b. pharynx, c. trachea).

4. The major muscle involved with the process of breathing is the (a. diaphragm, b. abdominal rectus, c. teres major).

5. Tiny hair-like structures that aid in keeping the air passages clean are called (a. flagella, b. cilia, c. villi).

6. The trachea is located (a. beside, b. in front of, c. in back of) the esophagus.

7. The structure that produces the sound of the voice as air passes through it is the (a. epiglottis, b. pharynx, c. larynx).

8. The lungs are located in the (a. thoracic, b. abdominal, c. pelvic) cavity of the body.

9. The trachea branches into structures called (a. portals, b. lobes, c. bronchial tubes).

10. A very serious disease of the lungs in which the lungs lose their elasticity is (a. cirrhosis, b. emphysema, c. edema).

11. Difficult or labored breathing is referred to as (a. dyspnea, b. hyperventilation, c. apnea).

12. The volume of air breathed out after a normal inspiration is called the (a. tidal volume, b. reserve volume, c. residual volume).

13. Each nasal cavity of the nose is divided into (a. two, b. three, c. four) passageways to aid in warming, cleaning, and humidifying incoming air.

14. A term used in association with the lungs is (a. pulmonary, b. hepatic, c. cardiac).

15. The left lung has two and the right lung has three (a. portal systems, b. lobes, c. cavities).

WORKSHEET 5-83

Cellular Respiration: Word Scramble

Read each clue and then unscramble the words in parentheses, writing the answers in the blanks provided.

1. During cellular respiration, the cell releases energy from this material.
   (LCUOSEG) __ __ __ __ __ __ __
                         5   24

2. The released energy from cellular respiration is stored in this substance.
   (TAP) __ __ __
            10 17

3. The type of cellular respiration that requires oxygen
   (RBIAOCE) __ __ __ __ __ __ __
                      22 15   12

4. The cell structure in which most of the reactions of cellular respiration occur
   (NOTMODIOHCIRN) __ __ __ __ __ __ __ __ __ __ __ __ __
                            26   14   8   4

5. In addition to energy, the other two products of cellular respiration
   (RBAONC) (XIDIODE) __ __ __ __ __ __ __ __ __ __ __ __ (AWTRE) __ __ __ __
                                31   27 19

6. The type of cellular respiration that does not require oxygen
   (NAEBACORI) __ __ __ __ __ __ __ __ __
                1 25   30

7. The first of the two stages of cellular respiration
   (LCYOGLYISS) __ __ __ __ __ __ __ __ __ __
                                         29

8. The second of the two stages of cellular respiration
   (TICIRC) (CDAI) (YCLCE) __ __ __ __ __ __ __ __ __ __ __ __ __
                            7 28   18   2

9. What ATP is the primary source of for the cell
   (NEREGY) __ __ __ __ __ __ __ __ __
                   21 11

10. Cells in the human body that are capable of anaerobic respiration
    (USMLCE) __ __ __ __ __ __
             23   6   9

11. Another name for the citric acid cycle
    (REKBS) (YLCCE) __ __ __ __ __ __ __ __ __ __
                     3   16   20

12. As the energy of ATP is used by the body, the ATP becomes what compound?
    (DAP) __ __ __
                 13

To answer the following, select the appropriate numbered letters from above and write them on the dotted lines to complete each statement:

a. The full name of ATP is

$\overline{}_{1}\ \overline{}_{2}\ \overline{}_{3}\ \overline{}_{4}\ \overline{}_{5}\ \overline{}_{6}\ \overline{}_{7}\ \overline{}_{8}\ \overline{}_{9}$ $\overline{}_{10}\ \overline{}_{11}\ \overline{}_{12}\ \overline{}_{13}\ \overline{}_{14}\ \overline{}_{15}\ \overline{}_{16}\ \overline{}_{17}$ H $\overline{}_{18}\ \overline{}_{19}\ \overline{}_{20}$.

b. The type of anaerobic respiration used by yeast in bread making is

F $\overline{}_{21}\ \overline{}_{22}\ \overline{}_{23}\ \overline{}_{24}\ \overline{}_{25}\ \overline{}_{26}\ \overline{}_{27}\ \overline{}_{28}\ \overline{}_{29}\ \overline{}_{30}\ \overline{}_{31}$.

WORKSHEET 5-84

Modes of Reproduction: Matching

For each of the following groups of terms, connect each term in the left-hand column with its proper counterpart at the right by drawing a straight line between them.

Group 1:

a. asexual	a. Development of an unfertilized egg
b. regeneration	b. Exchange of nuclear material between two unicellular organisms
c. parthenogenesis	c. Section of a cell grows and separates from original cell
d. fission	d. Reproduction involving nonsexual structures
e. conjugation	e. Unicellular organism divides into two.
f. budding	f. Growing back of a missing part

Group 2:

a. yeast	a. Conjugation
b. Paramecium	b. Spores
c. Amoeba	c. Budding
d. molds	d. Regeneration
e. honeybees	e. Fission

Group 3:

a. gametes	a. Sexually fertilized egg
b. zygote	b. Asexual mode of reproduction of molds
c. stolon	c. Sexual reproductive organs of plants
d. hermaphrodite	d. Cells of sexual reproduction
e. flowers	e. Artificial vegetative propagation
f. grafting	f. Having both male and female sex organs

WORKSHEET 5-85

Female Reproductive System: Around the Square

Using the clues and answers provided, fill in the squares with the appropriate answers beginning with number 1 at the lower left and continuing clockwise. (Selected numbers around the square are provided as a guide.)

Clues:

1. Produces the female eggs
2. Another name for the breasts
3. Another name for female (or male) sex cells
4. The inner mucous membrane lining the uterus
5. Tubes that transport the egg from the ovaries to the uterus
6. Another name for the female reproductive organs
7. Canal leading from the vagina to the uterus
8. Formal name for the female egg
9. Monthly discharge from the uterus
10. Lower part of the birth canal leading to outside of the body
11. Another name for the womb
12. Name of the monthly cycle that produces eggs

Answers:

endometrium	gonads
fimbriae	uterus
morula	areola
ovaries	mammary glands
vagina	gametes
ovarian	paraurethral
menstruation	ovum
Fallopian	cervical

Name: _____   Date: _____

WORKSHEET 5-86

Male Reproductive System: Around the Square

Using the clues and answers provided, fill in the squares with the appropriate answers beginning with number 1 at the lower left and continuing clockwise. (Selected numbers around the square are provided as a guide.)

Clues:

1. Secretes an alkaline fluid that is part of the seminal fluid
2. Secretes and provides ducts for seminal fluid
3. Also secretes an alkaline fluid that is part of the seminal fluid
4. Ducts that carry seminal fluid through the prostate gland to the urethra
5. Formal name for the reproductive fluids
6. Collective name for the male sex cells
7. Duct that carries seminal fluid from the epididymis to the ejaculatory ducts
8. Tube that leads to outside of body for reproductive and urinary systems
9. Another name for the reproductive organs
10. Formal name for the male sex cell
11. Structure in which the male sex cells are produced
12. Structures in the sperm that provide energy

Answers:

testes	prostate
gonads	mitochondria
epididymis	bulbourethral
vas deferens	gametes
ejaculatory	spermatozooan
urethra	lobule
semina	scrotum
Cowper's	meiosis

WORKSHEET 5-87

Body Senses: Matching

For each of the following groups of terms, connect each term in the left-hand column with its proper counterpart at the right by drawing a straight line between them.

Group 1:

a. mechanoreceptors   a. Pain receptors in the skin

b. cornea   b. Detect tension in muscles

c. free nerve endings   c. Touch receptors in skin

d. optic nerve   d. Detect pressure and vibrations

e. proprioceptors   e. Focuses light rays in eye

f. Meissner's corpuscles f. Carries nerve impulses from eye to brain

Group 2:

a. chemoreceptors   a. Amplify sound waves in ear

b. olfactory   b. Detect chemicals in substances

c. gustatory   c. Refers to sense of taste

d. semicircular canals   d. Pressure receptors in skin

e. ossicles   e. Refers to sense of smell

f. Pacinian corpuscles   f. Deals with body equilibrium

Group 3:

a. Organ of Corti   a. Tips of sensory nerve cells

b. papillae   b. Converts sound waves into nerve impulses in ear

c. receptors   c. Gives shape to back of eyeball

d. extrinsic   d. Receptive to light

e. vitreous gel   e. Location of taste buds

f. photoreceptors   f. Muscles that move eyeball

Name: _____   Date: _____

The Skin: Word Pyramid

Read the following clues, select the proper terms, and fill in the eight levels of the pyramid in any sequence. The terms for each level of the pyramid can be written in any order. The number of terms for each level is as follows: Level 1 = 3; Level 2 = 2; Level 3 = 1; Level 4 = 2; Level 5 = 1; Level 6 = 2; Levels 7 and 8 = 1 each.

LEVEL #

4 Material that gives skin color

1 Glands located in the external ear that secrete waxy material

2 Another name for the skin

8 Skin receptors sensitive to pain

1 Name of system that comprises the skin and its related
   structures

7 Skin glands that secrete an oily substance

3 Type of epithelial cells that make up epidermis of skin

LEVEL #

1 Fibers that give skin strength

5 Outer layer of dead skin cells

6 Most numerous of all skin glands

2 Layer of skin cells above the dermis

6 Skin receptors sensitive to touch

4 Fold of skin covering the root of the nail

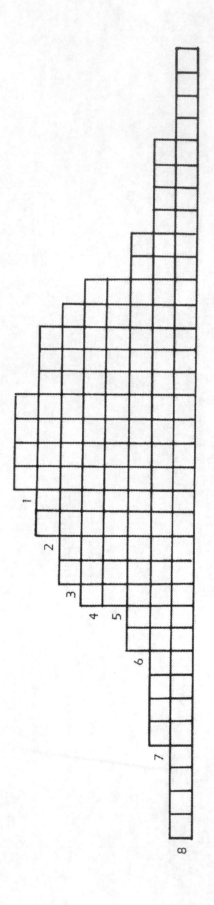

Terms:

cutaneous membrane   pain   integumentary   sweat   ceruminous   cuticle

stratified squamous   sebaceous   melanin   Ruffini's   collagenous   epidermis

stratum corneum

Name: _____   Date: _____

WORKSHEET 5-89

Plant Stems: Word Scramble

Unscramble the underlined word in each sentence and write the correct word on the dotted line.

1. One function of a stem is that of <u>ORTPPSU</u> __ __ __ __ __ __ $\overline{12}$.

2. Another function of a stem is that of <u>PROTSNTAR</u> __ __ __ __ __ __ __ __ __ .

3. The <u>ASVUCARL NDBLUE</u> __ __ __ __ __ __ __ __ __ __ __ __ contains both
   $\overline{1}$ $\overline{6}$ $\overline{13}$
   xylem and phloem tissues.

4. One type of a stem is called a <u>CODIT</u> __ __ __ __ __ .
   $\overline{15}$

5. Another type of stem is called a <u>NOOMCTO</u> __ __ __ __ __ __ __ .
   $\overline{2}$

6. Another name for a nonwoody stem is <u>REBHAECUOS</u> __ __ __ __ __ __ __ __ __ __ .
   $\overline{10}$

7. The outer layer of cells around some plant stems is called the <u>PDERISMEI</u> __ __ __ __
   __ __ __ __ .
   $\overline{19}$

8. The <u>IHTP</u> __ __ __ __ comprises the center of a dicot stem.
   $\overline{17}$

9. Tissue in which vascular bundles are found in a monocot stem is called the <u>OURGDN</u>
   <u>PRANCEHAMY</u> __ __ __ __ __ __ __ __ __ __ __ __ __ __ __ .

10. The wood of trees is comprised of <u>YLMEX</u> __ __ __ __ __ tissue.
    $\overline{18}$

11. The age of a tree can be determined by counting the number of <u>UNANAL NIGSR</u> __ __
    $\overline{8}$
    __ __ __ __ __ __ __ __ in its stem.
    $\overline{16}$ $\overline{7}$

12. The stems of some plants are adapted to carry on the process of <u>POOYHTTHSISENS</u>
    __ __ __ __ __ __ __ __ __ __ __ __ __ .
    $\overline{9}$ $\overline{14}$ $\overline{5}$

13. Stems that grow horizontally are called <u>OTLOSNS</u> __ __ __ __ __ __ __ .
    $\overline{3}$

14. A potato is actually a modified stem called a <u>BUERT</u> __ __ __ __ __ .
    $\overline{11}$

15. <u>ZOMIHRES</u> __ __ __ __ __ __ __ __ are another name for underground stems.
    $\overline{4}$

To answer the following, select the appropriate numbered letters above and write them in the spaces that complete each statement.

a.   The stem of a cactus plant is adapted for __ __ __ __ __ __ __   W __ __ __ __.

                                            1 2   3   4   5   6   7   8   9   10 11

b.   Stems that are adapted for climbing have specialized structures called

     __ __ __ __ __ __ __ __.

   12 13 14 15 16 17 18 19

Name: _____   **Date:** _____

WORKSHEET 5-90

Systems of the Body: Name That Category

Read the hints and determine the proper category to which each belongs. Write the category in the appropriate box. Categories can be found in the list at the bottom of the worksheet.

1. Body support and protection of internal organs
2. Body movements, posture, and heat production
3. Communication throughout the body via neurons
4. Transportation of materials to and from body cells
5. Breathing and cellular gas exchange

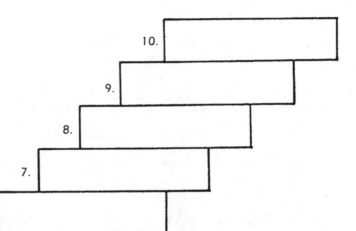

6. Collection and secretion of liquid body wastes
7. Production of male and female gametes
8. Communication throughout the body via hormones
9. Preparing ingested food materials for distribution to body cells
10. Defense and protection against disease organisms

Categories:

immune	skeletal	endocrine	reproductive	muscular
digestive	nervous	urinary	circulatory	respiratory

WORKSHEET 5-91

Teeth: True and False

Carefully read each statement. If the boldface word(s) makes the statement true, write a T in the blank; if false, write an F. Correct the false statements by crossing out the boldfaced word(s) and writing the correct word above it.

_____ 1. The main function of toothpaste **is** to serve as a breath freshener and perhaps a source of fluoride.

_____ 2. To effectively control dental plaque, the teeth should receive at least one daily brushing for a minimum of **two** minutes.

_____ 3. The function of fluoride is to strengthen **tooth enamel.**

_____ 4. A **hard**-bristle toothbrush is considered the most ideal for cleaning teeth.

_____ 5. It is actually the **toothbrush** that cleans the teeth rather than the toothpaste.

_____ 6. **Plaque** contains millions of living bacteria.

_____ 7. Plaque can be effectively controlled by flossing at least every **other** day.

_____ 8. Tartar is actually hardened **plaque** that has not been removed from the teeth.

_____ 9. A root canal **is** a painful dental procedure.

_____ 10. It is **normal** for the gums to bleed when brushing and/or flossing.

_____ 11. It is **normal** to expect some teeth to loosen as the years go by.

_____ 12. An **abscessed** tooth is one with an infected pulp cavity.

_____ 13. Using a water irrigation device **is** as effective as flossing in removing plaque between teeth.

_____ 14. The major cause of tooth loss during the adult years is **periodontal disease.**

_____ 15. An **endodontist** is a specialist in straightening teeth.

_____ 16. If a tooth is extracted, it is generally **okay** to leave the remaining gap.

_____ 17. **Malocclusion** is a dental problem characterized by the upper and lower sets of teeth not coming together properly in a bite.

_____ 18. **Discolored teeth** can be treated by a dental procedure called bonding.

Name: _____   **Date:** _____

General Biology Terms: Word Pyramid

Read the following clues, select the proper terms, and fill in the eight levels of the pyramid in any sequence. The terms for each level of the pyramid can be written in any order. The number of terms for each level is as follows: Level 1 = 3; Levels 2, 3, 4, 5, 6 = 2; Levels 7, 8 = 1.

LEVEL #

6	Color receptors of the eye
2	The division of the nervous system that controls heart rate
8	The basic unit of life
1	Opening in a leaf for water vapor and other gases
1	Individual filaments of some fungi organisms
7	An involuntary reaction
3	Phylum for oysters and clams
4	Has a relatively constant body temperature
6	Platelets help the blood to do this.
1	Connects the left and right cerebral hemispheres of brain
2	Obtains energy from carbon compounds
3	Movement of molecules from higher to lower concentration
5	Deals with organisms and their relationship to environment
5	An organism's role in the community
4	Female egg cell

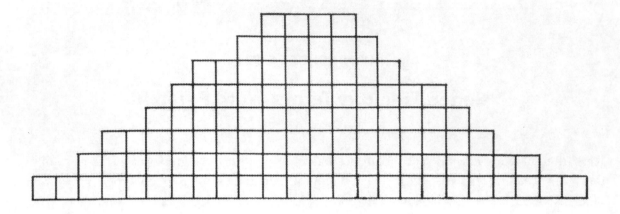

Terms:

corpus callosum	niche	cell	cones	Mollusca
heterotroph	reflex	ovum	ecology	stoma
homeotherm	clot	hypha	diffusion	autonomic

WORKSHEET 5-93

General Biology Terms: Word Pyramid

Read the following sentences, select the proper term, and fill in the seven levels of the pyramid in any sequence. The terms for each level of the pyramid can be written in any order. The number of terms for each level is as follows: Level 1 = 4; Level 2 = 3; Level 3 = 2; Level 4 = 2; Levels 5, 6, 7 = 1 each.

LEVEL #

7	Regulates amount of light entering eye
4	Major artery leaving the heart
2	Refers to the heart
1	Brief period of learning during early part of some animals' lives
6	Cell organelle involved in photosynthesis
5	Young are carried in an external pouch.
1	First of the two terms used in a scientific name
4	Digestive enzyme found in gastric juice
3	Third stage of mitosis
1	Another name for a nerve cell
2	Carries oxygen-rich blood
3	Can activate the body's immune system
2	Material comprises hard exoskeleton of some animals

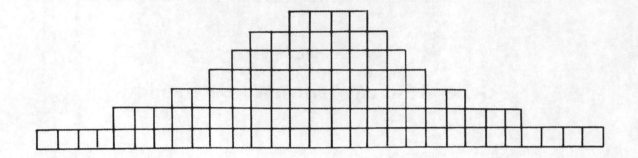

Terms:

aorta	iris	antigen	cardiac	plastid
lysosome	marsupial	neuron	pepsin	imprinting
chitin	anaphase	artery	arteries	vein

WORKSHEET 5-94

General Biology Terms: Fill in the Blanks

Select the proper term listed below and fill in the blanks of the sentences that follow.

1. The _____ is the most general or broadest classification group.
2. The outer layer of skin is the _____.
3. _____ can be caused by not having enough iron in the red blood cells.
4. The term _____ refers to the lungs.
5. A substance injected into the body containing weakened or nonliving pathogens is called a _____.
6. A _____ is a fertilized egg.
7. A _____ is a group of organs working together.
8. _____ refers to the ability of an organism to regrow lost body parts.
9. Bones are attached to bones by means of _____.
10. In addition to blood, _____ also circulates throughout the body in its own system of vessels.
11. A _____ is a seed that has only one cotyledon.
12. Organelles in the cell that are often called "powerhouses of the cell" are _____.
13. The _____ is the muscle used for breathing.
14. The substance in plants needed for photosynthesis is _____.
15. _____ are plants that die after one growing season.
16. The _____ stores food in the earthworm.
17. Molds and yeasts are _____ organisms.
18. Plant cell walls contain _____.
19. An animal that uses both plant and animal material for food is a _____.
20. An _____, produced by molds, is used to fight disease-causing germs.

Terms:

pulmonary	lymph	mitochondria	crop	antibiotic
anemia	fungus	omnivore	diaphragm	system
ligaments	kingdom	zygote	chlorophyll	vaccine
cellulose	regeneration	monocot	epidermis	annuals

Name: _____   **Date:** _____

WORKSHEET 5-95

General Biology Terms: Fill in the Blanks

Select the proper term listed below and fill in the blanks of the sentences that follow:

1. On a woody stem the _____ is the outer layer.
2. As an embryo develops it goes through a "hollow ball" stage called a _____.
3. The _____ surface is the front of an animal.
4. In birds the _____ is the organ in which food is ground up.
5. _____ means attached.
6. Genes arranged in sequence comprise the _____.
7. An organism that needs oxygen is said to be _____.
8. A _____ is a community of specific life forms.
9. One of the building blocks of the DNA molecule is _____.
10. A bone cell is called an _____.
11. _____ are folds of the cerebrum of the brain.
12. The back of an animal is the _____ surface.
13. A sheath in grass embryos is the _____.
14. _____ is soil enriched with decayed organisms.
15. One of the valves found in the heart is the _____.
16. The lungs contain _____, which are minute air sacs.
17. A sugar found in the structure of RNA is _____.
18. Organs of sensory detection in arthropods are called _____.
19. The _____ is an opening in the body of a sponge.
20. Starfish arms radiate out from its _____.

Answers:

biome	convolutions	chromosome	anterior	bark
gizzard	bicuspid	guanine	coleoptile	aerobic
alveoli	ribose	humus	dorsal	sessile
central disc	antennae	osteocyte	blastula	osculum

WORKSHEETS 5-96

Animal Tissues: Fill in the Blanks

Fill in the blanks of the statements below using proper terms selected from the list that follows the statements. (*Note:* Some terms will be used more than once.)

1. _____ muscle tissue is found only in the walls of the heart.
2. _____ tissue is the most widespread in the body.
3. _____ is another name for fat tissue.
4. Some tissues have structures called _____, which sweep the tissue clean.
5. The three types of muscle tissue are _____, _____, and _____.
6. Epithelial tissue which is organized of layers of cells is called _____.
7. _____ and _____ are the two types of involuntary muscle tissue.
8. _____ cells found in some types of tissue secrete mucous.
9. Support and protection is provided by _____ tissue.
10. Epithelial tissue which is only one cell layer thick is called _____.
11. Tissues are organizations of specialized _____.
12. _____ muscle tissue moves eyeballs and helps in swallowing.
13. _____ muscle tissue changes the diameter of blood vessels.
14. The tissue involved in transporting materials is _____.
15. The larynx (voicebox) is composed of _____.
16. The brain and spinal cord is composed of _____ tissue.
17. _____ tissues function in insulation and reserve food storage.
18. Cardiac and skeletal muscle tissue have cross markings called _____.
19. _____ tissue aids blood tissue in body defense against disease.
20. The _____ is an excellent example of stratified epithelial tissue.

Answers:

goblet	organ	skeletal	adipose	stratifications
cardiac	reticular	hyaline cartilage	nerve	elastic cartilage
systems	connective	simple squamous	visceral	striations
cilia	stratified	skin	cells	blood

WORKSHEET 5-97

Tracheophytes: Around the Square

Using the clues and answers provided, fill in the squares with the appropriate answers beginning with number 1 at the lower left and proceeding clockwise. (Selected numbers around the square have been provided as a guide.)

Clues:

1. The phylum name for tracheophytes
2. Because they have specialized vessels for transporting food and water, tracheophytes are called what?
3. Tracheophytes have these three structures, (three terms in answer)
4. Tracheophytes include these three plant groups, (three terms in answer)
5. Specialized vessels that carry water
6. Specialized vessels that carry food
7. During the sexual stage of a fern, this structure contains the egg- and sperm-forming tissues.
8. A tree that is a cone-bearing plant
9. A tree that is a flowering plant
10. Whereas cone-bearing and flowering plants produce seeds for reproduction, ferns produce what?

Terms:

spores	stems	ferns	flowering
phloem	vascular	Maple	leaves
prothallus	roots	xylem	Tracheophyta
cone-bearing	Pine	cells	seeds

3.

5

1.

10.

Name: _____   Date: _____

Cellular Transport Modes: Matching/Multiple Choice

Matching: Select the term in the right-hand column that best matches the definition in the left-hand column. Write the letter of the term in the blank.

_____ 1. The movement of molecules from a place of higher concentration to a place of lower concentration (may or may not be through a membrane)

_____ 2. The cell itself plays no active role in the passing of molecules through its membrane.

_____ 3. The process by which a portion of the cell membrane engulfs and takes in a small particle of solid material

_____ 4. The passing of water molecules only, through a membrane from a place of higher concentration to a place of lower concentration

_____ 5. Process by which substances are expelled from the cell through its membrane

_____ 6. The cell itself plays an active role in the passing of molecules through its membrane.

_____ 7. The process by which a portion of the cell membrane engulfs and takes in a small droplet of liquid

a. Active transpot

b. Osmosis

c. Pinocytosis

d. Diffusion

e. Passive transport

f. Exocytosis

g. Phagocytosis

Multiple Choice: Circle the letter of the correct answer to each of the following.

1. The process by which a white blood cell engulfs a bacterium is
   a. osmosis   b. phagocytosis   c. diffusion   d. pinocytosis.

2. A plant is wilting and you water it. The process by which the water passes through the plant's cell membranes to enter the cells is
   a. exocytosis   b. phagocytosis   c. osmosis   d. pinocytosis.

3. The processes which are active transport are
   a. diffusion and osmosis   b. phagocytosis, pinocytosis, and exocytosis.

4. The process by which molecules of sugar pass through an animal cell membrane to replenish the cell's lack of sugar is
   a. pinocytosis   b. exocytosis   c. osmosis   d. diffusion.

5. The process by which a cell gets rid of waste material by expelling it through its cell membrane is
   a. diffusion   b. osmosis   c. exocytosis   d. passive transport.

Name: _____   Date: _____

The Urinary System: Matching

Select the proper terms listed below to match the statements that follow. Write the letter of the term in the blank.

a. medulla	f. urethra	k. urinary bladder	p. antidiuretic
b. sugar	g. ureters	l. hemodialysis	q. micturition
c. right	h. renal	m. incontinence	r. left
d. cortex	i. filtration	n. nephron unit	s. Bowman's capsule
e. tubules	j. glomerulus	o. urea	t. blood transport

_____ 1. Carries urine from the kidneys to the urinary bladder

_____ 2. Term for the waste products of protein metabolism eliminated by the kidneys

_____ 3. The process by which the blood is cleansed by the kidneys

_____ 4. Stores urine prior to its being eliminated from the body

_____ 5. Microscopic units within the kidneys that aid in urine formation

_____ 6. Type of hormone that slows down water loss through the kidneys

_____ 7. Process by which the blood is cleansed by an artificial kidney machine

_____ 8. Part of a nephron unit, this structure consists of a network of blood capillaries.

_____ 9. The terms used for arteries and veins which serve the kidneys

_____ 10. The structure which eliminates urine to the outside of the body

_____ 11. The name for the outer layer of the kidney

_____ 12. Part of the nephron unit, this structure surrounds the glomerulus.

_____ 13. The kidney that is lower than the other in the body

_____ 14. The name for the inner layer of the kidneys

_____ 15. Structures that transport urine from the kidneys to the ureters

_____ 16. Substance which, when found in urine, suggests diabetes

_____ 17. Inability to control the release of urine from the body

_____ 18. The clinical term for the process of urinating

WORKSHEET 5-100

Vitamins and Minerals: Fill in the Blanks

Select the proper answer listed below to fill in the blanks of the sentences that follow.

1. _____ are organic molecules, and _____ are inorganic molecules, both of which are essential for proper cell metabolism.

2. The B and C vitamins are _____ soluble, which allows for excess amounts in the body to be excreted in the urine.

3. Vitamins A, D, E, and K are _____ soluble; thus excess amounts in the body can build up to toxic levels in body tissues.

4. The mineral _____ is a major component of bone and teeth.

5. The mineral _____, along with calcium, is important for normal muscle and nerve function.

6. The mineral _____ is needed for the production of the red blood cells.

7. The mineral _____ strengthens the enamel of teeth, helping to protect against cavities.

8. Minerals needed by the body in small amounts are called _____.

9. Vitamin _____ is essential for the normal clotting of blood.

10. A lack of vitamin _____ can lead to a type of anemia.

11. A lack of vitamin _____ can lead to scurvy.

12. Vitamin _____ is needed for normal bone structure.

13. A lack of vitamin _____ can lead to structurally unsound red blood cells.

14. An excellent dietary source of vitamin C is _____.

15. An excellent dietary source of vitamin B is _____.

16. Vitamin D is added to the food product _____.

17. The mineral iodine is added to the food product _____.

Answers:

D_3	fat	trace minerals	salt	vitamins
milk	minerals	K	calcium	C
water	whole grains	magnesium	cobalt	citrus fruits
B_{12}	E	fluorine	minute	fiber

Answers to the Sponge Activities:

5-1: Algae: Word Scramble

1. brown; 2. Chlorophyta; 3. Volvox; 4. Ulothrix; 5. Chlorella; 6. Spirogyra; 7. Desmid; 8. Phaeophyta; 9. Kelp; 10. Rhodophyta; 11. Diatoms; 12. Chlorophyll; 13. Phytoplankton; a. conjugation; b. algin, carrageenan.

5-2: Amphibians: Fill in the Blanks

1. water; 2. carnivores; 3. Chordata; 4. cold-blooded; 5. three; 6. external; 7. metamorphosis; 8. double life; 9. frogs; 10. tail; 11. Bufo; 12. salamanders; 13. larval; 14. jelly-like; 15. toads; 16. gills; 17. Rana Pipiens; 18. scales; 19. claws.

5-3: Animal Development: Matching

Group 1: a-b; b-f; c-a; d-c; e-e; f-d.
Group 2: a-f; b-a; c-b; d-c; e-d; f-e.
Group 3: a-c; b-d; c-f; d-a; e-b; f-e.

5-4: Animal Groups: Matching

1. M; 2. A; 3. I; 4. R; 5. F; 6. M; 7. M; 8. M; 9. F; 10. A; 11. I; 12. M; 13. R; 14. M; 15. A; 16. R; 17. M; 18. R; 19. R; 20. M; 21. M; 22. M; 23. R; 24. R; 25. M; 26. B; 27. I; 28. B; 29. M; 30. B; 31. M; 32. M; 33. M; 34. F; 35. F; 36. M; 37. M; 38. M; 39. M; 40. I; 41. I; 42. I; 43. I; 44. I; 45. B; 46. M; 47. B; 48. I; 49. M; 50. B; 51. I; 52. I; 53. M; 54. I; 55. I; 56. I; 57. I; 58. F; 59. F; 60. R.

5-5: Animal Phyla: Name That Category

1. Arthropoda; 2. Porifera; 3. Chordata; 4. Echinodermata; 5. Annelida; 6. Cnidaria; 7. Hemichordata; 8. Platyhelminthes; 9. Mollusca; 10. Aschelminthes.

5-6: Animal Phyla Members: Matching

1. E; 2. D; 3. H; 4. G; 5. I; 6. J; 7. B; 8. F; 9. C; 10. D; 11. D; 12. D; 13. B; 14. F; 15. J; 16. D; 17. B; 18. D; 19. H; 20. G; 21. J; 22. D; 23. I; 24. D; 25. B; 26. H; 27. J; 28. G; 29. G; 30. C; 31. C; 32. A; 33. D; 34. F; 35. C; 36. J; 37. D; 38. C; 39. D; 40. C; 41. J; 42. J; 43. H; 44. C; 45. C; 46. D; 47. J; 48. I; 49. F; 50. D; 51. D.

5-7: Facts of Biology: Letter Search

Hint 1: beaver; Hint 2: upper incisors.

5-8: Facts of Biology: Letter Search

Hint 1: middle ear; Hint 2: masseter.

5-9: Facts of Biology: Letter Search

Hint 1: neurons; Hint 2: sneezing.

5-10: Facts of Biology: Letter Search

Hint 1: pituitary; Hint 2: crying.

5-11: Facts of Biology: Letter Search

Hint 1: homeostasis; Hint 2: femur.

5-12: Facts of Biology: Letter Search

Hint 1: triglyceride; Hint 2: cholesterol.

5-13: Facts of Biology: Letter Search

Hint 1: blue whale; Hint 2: sequoia.

5-14: Facts of Biology: Letter Search

Hint 1: cardiovascular; Hint 2: liver.

5-15: Facts of Biology: Letter Search

Hint 1: osculation; Hint 2: osulation

5-16: General Biology: Around the Square

Answers clockwise: study of life; Euglena; metric system; microscopes; Louis Pasteur; ribosome; chromosome; Linnaeus; stamen; protozoa; cell.

5-17: General Biology: Name That Category

1. phylum names; 2. microscope types; 3. photosynthesis; 4. Mendel and genetics; 5. bacteria shapes; 6. locomotion structures; 7. fungi; 8. conifers; 9. asexual reproduction; 10. seed parts; 11. worms; 12. insects.

5-18: General Biology: Name That Category

1. food groups; 2. DNA parts; 3. mitosis stages; 4. cell parts; 5. skeletal bones; 6. microscope parts; 7. vitamins; 8. green algae; 9. kingdoms; 10. cell transport; 11. nonvascular plants; 12. flower parts.

5-19: General Biology: True and False

1. F (producers; 2. F (lipids); 3. F (osmosis); 4. T; 5. F (Gregor Mendel); 6. F (East Africa); 7. F (bacteria; 8. T; 9. T; 10. T; 11. F (autotroph); 12. T; 13. F (all); 14. T; 15. T; 16. T; 17. F (*Homo sapiens*); 18. T; 19. F (transpiration); 20. F (biosphere).

5-20: Biomes: Matching

1. I; 2. G; 3. D; 4. J; 5. A; 6. F; 7. K; 8. C; 9. M; 10. B; 11. N; 12. H; 13. L; 14. E; 15. 0.

5-21: Body Movements and Positions: Name That Category

1. abduction; 2. rotation; 3. flexion; 4. pronation; 5. anatomical position; 6. supinated position; 7. extension; 8. supination; 9. adduction; 10. pronated position.

5-22: Bones of the Body: Fill in the Blanks

1. humerus; 2. patella; 3. femur; 4. cranium; 5. mandible; 6. carpals; 7. phlanges; 8. frontal; 9. innominate; 10. tarsals; 11. clavicle; 12. sternum; 13. radius; 14. scapula; 15. vertebrae; 16. tibia; 17. maxilla; 18. malar (zygomatic); 19. fibula; 20. ulna.

5-23: Brain and Brain Stem Functions: Matching

1. C; 2. F; 3. L; 4. K; 5. J; 6. H; 7. E; 8. G; 9. K; 10. C; 11. B; 12. M; 13. F; 14. D; 15. I; 16. A.

5-24: The Brain, Brain Stem, and Spinal Cord: Color Coding

No answers required.

5-25: Bryophytes: Matching

1. D; 2. H; 3. J; 4. K; 5. L; 6. N; 7. Q; 8. B; 9. O; 10. E; 11. P; 12. M; 13. F; 14. G; 15. A; 16. I; 17. C.

5-26: Cell Parts and Functions: Fill in the Blanks

1. cytoplasm; 2. nucleus; 3. chloroplast; 4. ribosomes; 5. cell membrane; 6. mitochondria; 7. nuclear membrane; 8. lysosomes; 9. cell wall; 10. vacuoles; 11. centrioles; 12. Golgi complex; 13. nucleolus; 14. endoplasmic reticulum; 15. chromatin.

5-27: Cell Structure: Color Coding

No answers required.

5-28: Basic Chemistry: Message Square

Hint: chemical reaction; Research Questions: 1. physical; 2. in a mixture the substances are NOT chemically combined; 3. one substance dissolved in another substance resulting in a mixture that is homogeneous throughout; 4. the substance being dissolved is the solute; the substance doing the dissolving is the solvent; 5. a mixture; the suspended substance is not dissolved.

5-29: Basic Chemistry: Message Square

Hint: chemical formula; Research Questions: the basic building block of an element; 2. particles of matter comprised of atoms; 3. a substance that cannot be broken down into simpler substances; 4. 103; 5. an electrically charged atom; 6. an acid has an excess of hydrogen (H) ions and a base has an excess of hydroxyl (OH) ions.

5-30: The Circulatory System: Sentence Correction

1. erythrocytes; 2. cholesterol; 3. hematocrit; 4. carry food and oxygen; 5. aorta; 6. thrombus; 7. thrombocytes; 8. pulse; 9. leukocytes; 10. plasma; 11. clot blood; 12. protect the body against disease; 13. vena cava; 14. elastic walls; 15. differential count; 16. valves; 17. myocardial infarction; 18. pulmonary; 19. sinoatrial node; 20. electrocardiogram.

5-31: Classification of Monerans, Protists, and Fungi: Matching

Group 1: a-e; b-a; c-b; d-c; e-d.
Group 2: a-e; b-a; c-b; d-f; e-c; f-d.

Group 3: a-d; b-f; c-b; d-c; e-a; f-e.

5-32: Classification: Message Square

Hint: plant, protist, animal; Research Questions: 1. for organisms that could not be classified as true plants or true animals; 2. Fungi Kingdom, which includes molds, yeasts, and mushrooms; Moneran Kingdom, which includes bacteria and blue-green algae; 3. Carolus Linnaeus.

5-33: Classification: Crossword

Across: 2. eukaryotic; 4. species; 6. taxonomy; 7. Linnaeus; 9. Latin; 10. order; 12. insects; 14. kingdom; 15. protista.

Down: 1. binomial nomenclature; 5. genus; 8. autotrophic; 15. fungi; 16. heterotrophic. Up: 3. Canis familiaris; 11. Felis domesticus; 13. Prokaryotic.

5-34: The Cranial Nerves: Fill in the Blanks

vision = optic (#2); face and head sensations = trigeminal (#5); eyeball movements = trochlear (#4) or abducens (#6); movements of the tongue = hypoglossal (#12); smell = olfactory (#1); taste, saliva secretion, facial muscle movements = facial (#7); eyeball movements, pupil size regulation = oculomotor (#3); turning head = accessory (#11); eyeball movements = abducens (#6) or trochlear (#4); taste perception, swallowing = glossopharyngeal (#9); hearing, balance = acoustic (#8); slows heart rate, movements of internal organs such as stomach and larynx = vagus (#10).

5-35: Digestion: Word Scramble

1. fiber; 2. fatty acids; 3. amino acids; 4. glucose; 5. sugars; 6. enzymes; 7. peristalsis; 8. bile; 9. small intestine; 10. lipases; 11. salivary; 12. jejunum; 13. ascending colon; 14. esophagus; 15. chyme; a. Hydrolysis.

5-36: Organs of the Digestive System: Matching

1.d; 2.a; 3.c; 4.g; 5.h; 6.n; 7.q; 8.j; 9.e; 10.r; 11.p; 12.k; 13.f; 14.m; 15.b; 16.o; 17.i; 18.l.

5-37: The Structure of DNA: Color Coding/Fill in the Blanks

1. nitrogen; 2. guanine, thymine; 3. sugar, phosphate; 4. chromosomes; 5. zipper; 6. James Watson, Francis Crick; 7. genes.

5-38: Directional Terms Used in Association with the Body: Crossword

Down: 1. anterior; 5. transverse; 9. frontal; 11. ventral; 16. proximal.
Across: 2. right; 4. sagittal; 6. superior; 8. posterior; 10. love; 12. dorsal; 13. nothing; 15. distal; 17. medial.
Up: 3. lateral; 7. inferior; 14. good.

5-39: Diseases: Name That Category

1. AIDS; 2. tetanus; 3. diabetes; 4. atherosclerosis; 5. myocardial infarction; 6. measles; 7. mononucleosis; 8. mumps; 9. whooping cough; 10. chicken pox; 11. influenza; 12. food poisoning.

5-40: Diseases: Name That Category

1. epithelial carcinoma; 2. diverticulosis; 3. tonsilitis; 4. glaucoma; 5. emphysema; 6. familial hypercholesterolemia; 7. anemia; 8. dermatitis; 9. strep throat; 10. cataract; 11. gastric ulcer; 12. ringworm.

5-41: Terms Associated with Disease-Causing Organisms: Matching

1.K; 2.R; 3.J; 4.A; 5.F; 6.N; 7.D; 8.L; 9.Q; 10.B; 11.M; 12.C; 13.H; 14.I; 15.E; 16.P; 17. O; 18.G.

5-42: The Ear: Fill in the Blanks

1. three; 2. amplify and carry; 3. tympanic membrane; 4. cochlea; 5. ossicles; 6. malleus; 7. incus; 8. stapes; 9. semicircular canals; 10. Eustachian; 11. tinnitis; 12. auditory nerve; 13. auricle; 14. equalize; 15. sound waves.

5-43: Endocrine Gland Hormones: Matching

1.i; 2.k; 3.a; 4.h; 5.h; 6.b; 7.c; 8.f; 9.g; 10.c; 11.j; 12.c; 13.d; 14.a; 15.f; 16.e; 17.c; 18.c; 19.a; 20.d.

5-44: Functions of Endocrine Gland Hormones: Matching

1.e; 2.h; 3.l; 4.h; 5.n; 6.o; 7.c; 8.g; 9.f, r; 10.b; 11.m; 12.k; 13.j; 14.d; 15.i; 16.p; 17.q; 18.r.

5-45: Locations of Endocrine Glands: Message Square

Hint: Hormones; Research Questions: 1. a) top surface of kidneys; b) stomach lining; c) base of brain in cranial cavity; d) pelvic portion of abdominopelvic cavity; e) pancreas; f) neck; g) chest cavity; h) scrotum; i) uterus during pregnancy; j) neck behind thyroid; k) behind the third ventricle of the brain.

5-46: Environmental Adaptation: Message Square

Hint: physiological, behavioral; Research Questions: 1. one organism resembling another organism as a protective mechanism; 2. protection for the mimic; 3. the robber fly resembles the bumble bee, the latter ignored by many predators; 4. birds migrating to warmer regions during the winter; 5. a green plant growing toward light.

5-47: Environmental Adaptation: Message Square

Hint 1: chromatophores; Hint 2: cryptic coloration; Research Questions: 1. A display by an organism of a very noticeable color to ward off predators. Usually displayed by organisms that have an offensive taste or odor.; 2. To ward off predators by linking color appearance with offensive taste or odor.

5-48: Environment and Pollution: Matching

1.d; 2.i; 3.k; 4.a; 5.g; 6.h; 7.b; 8.f; 9.p; 10.r; 11.c; 12.o; 13.e; 14.n; 15.l; 16.j; 17.m; 18.q.

5-49: Enzymes: Word Scramble

1. protein; 2. chemical reaction; 3. substrate; 4. amylases; 5. proteases; 6. lipases; 7. lock and key; 8. coenzymes; 9. activation; 10. catalysts; 11. rate; 12. active site; 13. anabolism; 14. induced fit; a. luciferase.

5-50: Experimentation: Message Square

Hint: controlled; Research Questions: 1. When carrying out an experiment, the scientist attempts to CONTROL all conditions under which the experiment is conducted.; 2. control group, experimental group; 3. an idea or question asked by a scientist that can be tested; 4. a hypothesis that has been verified through testing; 5. information obtained through experimentation and observation; 6. analysis (of data).

5-51: Structure of the Eye: Labeling

1. cornea; 2. pupil; 3. iris; 4. suspensory ligaments; 5. lens; 6. aqueous humor; 7. sclera layer; 8. choroid layer; 9. retina layer; 10. optic disk (blind spot); 11. fovea; 12. optic nerve; 13. vitreous gel (body).

5-52: Eye Structure and Function: Anagrams

1. pupil; 2. fovea; 3. blind spot; 4. iris; 5. choroid; 6. vitreous body; 7. optic nerve; 8. aqueous humor; 9. conjunctiva; 10. sclera; 11. cornea; 12. retina; 13. suspensory ligaments; 14. lens; 15. rods, 16. cones.

5-53: Fields of Biology: Anagrams

1. botany; 2. zoology; 3. microbiology; 4. oceanography; 5. mycology; 6. entomology; 7. ichthyology; 8. physiology; 9. ecology; 10. nutritionist; a. anatomy; b. cytology; c. genetics.

5-54: Fish: True and False

1.T; 2.F, zoologist; 3.F, vertebrates; 4.T; 5.F, cold; 6.F, soft-shelled; 7.T; 8.F, two; 9.T; 10.T; 11.F, dorsal; 12.F, operculum; 13.F, ovoviparous; 14.T; 15.T; 16.F, swim bladder; 17.F, lateral line; 18.T; 19.T; 20.F, mammals.

5-55: The Anatomy of a Fish: Color Coding/Labeling

No answers required.

5-56: Structure of a Typical Flower: Labeling/Fill in the Blanks

Label: 1.anther; 2.stamen; 3.filament; 4.sepals; 5.petals; 6.stigma; 7.style; 8.pistil; 9.ovary. Fill in the blanks: 1. ovary; 2. stigma; 3. sepals; 4. filament; 5. style; 6. petals; 7. anther; 8. pistil; 9. stamen.

5-57: The Food Chain: Crossword

Across: 1. energy; 3. sun; 4. consumers; 6. animals; 8. herbivores; 10. carnivores.
Down: 2. producers; 5. plants; 7. first level; 9. omnivores.

5-58: The Anatomy of a Frog: Color Coding/Labeling

No answers required.

5-59: Genetics: Matching

Group 1: a-c; b-d; c-f; d-b; e-e; f-a.
Group 2: a-b; b-d; c-e; d-a; e-c; f-f.
Group 3: a-d; b-e; c-f; d-g; e-c; f-b.

5-60: Genetics: Sentence Correction

1. heredity; 2. dominant; 3. phenotype; 4. genotype; 5. replication; 6. recessive; 7. meiosis; 8. DNA; 9. mutation; 10. zygote; 11. sex chromosomes; 12. genes; 13. 46; 14. sex-linked trait; 15. color blindness; 16. 23; 17. albino.

5-61: Genetic Diseases and Diagnosis

1. red-green; 2. hemophilia; 3. Klinefelter's syndrome; 4. Down's syndrome; 5. Turner's syndrome; 6. Tay-Sachs; 7. sickle-cell anemia; 8. phenylketonuria; 9. mutation; 10. amniocentesis; 11. ultrasound; a. fetoscopy; b. mutagen.

5-62: The Human Heart: Around the Square

Answers clockwise: chambers; vena cava; ventricles; atria; mitral; tricuspid; aorta; pulmonary; apex; oxygenated; pulmonary; cardiac; smoking; coronary.

5-63: Immune System: Word Scramble

1. white; 2. antigen; 3. antibody; 4. thymus; 5. interferon; 6. rejection; 7. cyclosporine; 8. lymphocyte; 9. phagocytosis; 10. AIDS; 11. vaccine; a. active; b. passive; c. autoimmune.

5-64: Laboratory Equipment: Crossword

Across: 2. terrarium; 4. beaker; 6. scalpel; 8. stethoscope; 10. torso; 11. computer; 13. magnetic.
Down: 1. Petri; 3. pipet; 5. forceps; 9. oscilloscope; 14. autoclave.
Up: 7. stains; 12. microprojector.

5-65: Laboratory Equipment: Message Square

Hint 1: goggles, apron: Hint 2: spectroscope; Research Questions: 1. It produces a spectrum of the material that it is analyzing, for example a gas.; 2. Light from the star is passed through the spectroscope. By studying the spectrum, it can be determined what gases comprise the star.

5-66: Laboratory Equipment: Message Square

Hint: student and binocular; Research Questions: 1. The binocular microscope allows three-dimensional viewing of the specimen.; 2. scanning; 3. magnetism.

5-67: Internal Structures and Functions of a Leaf: Matching

1.g; 2.b; 3.a; 4.j; 5.h; 6.e; 7.d; 8.k; 9.f; 10.i; 11.c; 12.1.

5-68: Leaf Structure: Labeling

Part 1: 1. cutin; 2. upper epidermis; 3. palisade layer; 4. spongy layer; 5. xylem; 6. phloem; 7. lower epidermis; 8. stoma; 9. guard cell.
Part 2: 1. sinus; 2. lobe; 3. tip; 4. double-toothed; 5. leafstalk; 6. vein; 7. single-toothed; 8. midrib; 9. wavy-edged.

5-69: Deciduous Leaf Types and Arrangements: Matching

a. alternate compound; b. opposite compound; c. Scarlet oak; d. Weeping willow; e. Sugar maple; f. American elm; g. alternate simple; h. Sassafras; i. Sycamore; j. Tulip tree; k. opposite simple; 1. Gray birch; m. holly; n. mulberry; o. Flowering dogwood.

5-70: Medical Science Terms: Anagrams

1. diagnosis; 2. infection; 3. hypodermic; 4. prognosis; 5. immunity; 6. carcinogen; 7. biopsy; 8. benign; 9. thrombus; 10. hypertension; 11. insomnia; 12. syndrome; 13. incubation; 14. virus; 15. sterile; 16. ambulatory; 17. coronary; 18. orthopedics; 19. dermatology; 20. epidemic; 21. stroke; 22. gastric.

5-71: Meiosis: Message Square

Hint: male and female sex cells; Research Questions: 1. gametes; 2. The number of chromosomes in each of the resulting daughter cells has been reduced by one half that of the parent cell.; 3. a. 45; b. 23; 4. ova (eggs); 5. spermatozoa (sperm); 6. 46.

5-72: Mitosis: Matching

1.j; 2.g; 3.a; 4.b; 5.d; 6.c; 7.i; 8.k; 9.f; 10.h; 11.e.

5-73: Muscle Groups and Actions: Multiple Choice

1.a; 2.a; 3.b; 4.c; 5.c; 6.a; 7.a; 8.b; 9.a; 10.a; 11.c; 12.c; 13.a; 14.a; 15.c; 16.b.

5-74: Muscles of the Body: Fill in the Blanks

1. biceps; 2. triceps; 3. brachialis; 4. pectoralis major; 5. deltoid; 6. sternocleidomastoid; 7. trapezius; 8. latissimus dorsi; 9. gluteus maximus; 10. sartorius; 11. rectus femoris; 12. vastus lateralis; 13. vastus medialis; 14. quadriceps;

15. biceps femoris; 16. hamstrings; 17. gastrocnemius; 18. rectus abdominus; 19. frontalis; 20. masseter; 21. tibialis.

5-75: The Nervous System: True and False

1.T; 2.F, central; 3.F, cell body; 4.T; 5.F, meanings; 6.F, four; 7.F, cranial; 8.T; 9.F, to; 10.T; 11.T; 12.F, cerebro spinal fluid; 13.F, gray; 14.T; 15.T; 16.F, smell; 17.T; 18.T; 19.F, vertebrae; 20.F, dendrites.

5-76: Nutrition: Fill in the Blanks

1. nine; 2. saturated; 3. liver; 4. four; 5. fat; 6. exercise; 7. 150 mg; 8. vitamin; 9. generic; 10. greatest; 11. calories; 12. gain; 13. nutritionist; 14. proteins; 15. sugar; 16. fiber.

5-77: Phobias: Name That Category

1. claustrophobia; 2. hydrophobia; 3. xenophobia; 4. arachnophobia; 5. acrophobia; 6. nyctophobia; 7. necrophobia; 8. agoraphobia; 9. hemophobia; 10. zoophobia; 11. pyrophobia; 12. behavioral modification.

5-78: Photosynthesis: Around the Square

Answers clockwise: 1. respiration; 2. chlorophyll; 3. light, dark; 4. water, carbon dioxide; 5. oxygen, glucose; 6. grana; 7. palisade; 8. red, violet; 9. starch; 10. producers.

5-79: Plant Growth Responses: Word Scramble

1. stimulus; 2. tropism; 3. auxins, gibberellins; 4. geotropism; 5. phototropism; 6. photoperiodism; 7. chemotropism; 8. thigmotropism; 9. vernalization; 10. dormancy; 11. phytochrome; 12. florigen; a. nastic; b. plant physiologists.

5-80: Plants, The Higher: Message Square

Hint 1: vascular; Hint 2: xylem, phloem; Research Questions: 1. Tracheophyta; 2. cone bearing, flowering, ferns; 3. The xylem carries water and minerals from the roots up to the leaves. The phloem carries food throughout the plant.

5-81: Protozoans: Crossword

Across: 4. Protista; 5. zooplankton; 6. paramecium; 10. contractile vacuole; 11. flagellates; 12. Plasmodium.

Down: 1. amoeba; 2. consumer; 3. conjugation; 7. pseudopods; 8. ciliates; 9. nucleus.

5-82: Respiratory System: Multiple Choice

1.b; 2.b; 3.c; 4.a; 5.b; 6.b; 7.c; 8.a; 9.c; 10.b; 11.a; 12.a; 13.b; 14.a; 15.b.

5-83: Cellular Respiration: Word Scramble

1. glucose; 2. ATP; 3. aerobic; 4. mitochondrion; 5. carbon dioxide; 6. anaerobic; 7. glycolysis; 8. citric acid cycle; 9. energy; 10. muscle; 11. Krebs cycle; 12. ADP; a. adenosine triphosphate; b. fermentation.

5-84: Modes of Reproduction: Matching

Group 1: a-d; b-f; c-a; d-e; e-b; f-c.
Group 2: a-c; b-a; c-d; d-b; e-e.
Group 3: a-d; b-a; c-b; d-f; e-c; f-e.

5-85: Female Reproductive System: Around the Square

Answers clockwise: 1. ovaries; 2. mammary glands; 3. gametes; 4. endometrium; 5. Fallopian; 6. gonads; 7. cervical; 8. ovum; 9. menstruation; 10. vagina; 11. uterus 12. ovarian.

5-86: Male Reproductive System: Around the Square

Answers clockwise: 1. Cowper's; 2. epididymis; 3. prostate; 4. ejaculatory; 5. seminal; 6. gametes; 7. vas deferens; 8. urethra; 9. gonads; 10. spermatozooan; 11. testes; 12. mitochondria.

5-87: Body Senses: Matching

Group 1: a-d; b-e; c-a; d-f; e-b; f-c.
Group 2: a-b; b-e; c-c; d-f; e-a; f-d.
Group 3: a-b; b-e; c-a; d-f; e-c; f-d.

5-88: The Skin: Word Pyramid

Level 1: ceruminous, integumentary, collagenous.
Level 2: cutaneous membrane, epidermis.
Level 3: stratified squamous.
Level 4: melanin, cuticle.
Level 5: stratum corneum.

Level 6: sweat, Ruffini.
Level 7: sebaceous.
Level 8: pain.

5-89: Plant Stems: Word Scramble

1. support; 2. transport; 3. vascular bundle; 4. dicot; 5. monocot; 6. herbaceous; 7. epidermis; 8. pith; 9. ground parenchyma; 10. xylem; 11. annual rings; 12. photosynthesis; 13. stolons; 14. tuber; 15. rhizome; a. storing water; b. tendrils.

5-90: Systems of the Body: Name That Category

1. skeletal; 2. muscular; 3. nervous; 4. circulatory; 5. respiratory; 6. urinary; 7. reproductive; 8. endocrine; 9. digestive; 10. immune.

5-91: Teeth: True and False

1.T; 2.F, five; 3.T; 4.F, soft; 5.T; 6.T; 7.F, single; 8.T; 9.F, is not; 10.F, abnormal; 11.F, abnormal; 12.T; 13.F, is not; 14.T; 15.T; 16.F, not okay; 17.T; 18.T.

5-92: General Biology Terms: Word Pyramid

Level 1: stoma, hypha, corpus callosum.
Level 2: autonomic, heterotroph.
Level 3: Mollusca, diffusion.
Level 4: homeotherm, ovum.
Level 5: ecology, niche.
Level 6: cones, clot.
Level 7: reflex.
Level 8: cell.

5-93: General Biology Terms: Word Pyramid

Level 1: lysosome, imprinting, genus, neuron.
Level 2: cardiac, arteries, chitin.
Level 3: anaphase, antigen.
Level 4: aorta, pepsin.
Level 5: marsupial.
Level 6: plastid.
Level 7: iris.

5-94: General Biology Terms: Fill in the Blanks

1. kingdom; 2. epidermis; 3. anemia; 4. pulmonary; 5. vaccine; 6. zygote; 7. system; 8. regeneration; 9. ligaments; 10. lymph; 11. monocot; 12. mitochondria; 13. diaphragm; 14. chlorophyll; 15. annuals; 16. crop; 17. fungus; 18. cellulose; 19. omnivore; 20. antibiotic.

5-95: General Biology Terms: Fill in the Blanks

1. bark; 2. blastula; 3. anterior; 4. gizzard; 5. sessile; 6. chromosome; 7. aerobic; 8. biome; 9. guanine; 10. osteocyte; 11. convolutions; 12. dorsal; 13. coleoptile; 14. humus; 15. bicuspid; 16. alveoli; 17. ribose; 18. antennae; 19. osculum; 20. central disc.

5-96: Animal Tissues: Fill in the Blanks

1. cardiac; 2. connective; 3. adipose; 4. cilia; 5. cardiac, visceral, skeletal; 6. stratified; 7. cardiac, visceral; 8. goblet; 9. skeletal; 10. simple squamous; 11. cells; 12. skeletal; 13. visceral; 14. blood; 15. hyaline cartilage; 16. nerve; 17. adipose; 18. striations; 19. reticular; 20. skin.

5-97: Tracheophytes: Around the Square

Answers clockwise: 1. Tracheophyta; 2. vascular; 3. roots, stems, leaves; 4. ferns, cone-bearing, flowering; 5. xylem; 6. phloem; 7. prothallus; 8. pine; 9. maple; 10. spores.

5-98: Cellular Transport Modes: Matching/Multiple Choice

Matching:1.d; 2.e; 3.g; 4.b; 5.f; 6.a; 7.c.
Multiple choice: 1.a; 2.c; 3.b; 4.d; 5.c.

5-99: The Urinary System: Matching

1.g; 2.o; 3.i; 4.k; 5.n; 6.p; 7.l; 8.j; 9.h; 10.f; 11.d; 12.s; 13.c; 14.a; 15.e; 16.b; 17.m; 18.q.

5-100: Vitamins and Minerals: Fill in the Blanks

1. vitamins; 2. water; 3. fat; 4. calcium; 5. magnesium; 6. cobalt; 7. flourine; 8. trace; 9. K; 10. B_{12}; 11. C; 12. D_3; 13. E; 14. citrus fruits; 15. whole grains; 16. milk; 17. salt.

Michael F. Fleming

BIOLOGY TEACHER'S
Survival Guide

This unique resource is packed with novel and innovative ideas and activities you can put to use immediately to enliven and enrich your teaching of biology, streamline your classroom management, and free up your time to accomplish the many other tasks teachers constantly face.

For easy use, materials are printed in a big 8 ¼" × 11" lay-flat binding that opens flat for photocopying of evaluation forms and student activity sheets, and are organized into five distinct sections:

1. **Innovative Classroom Techniques for the Teacher** presents techniques to help you stimulate active student participation in the learning process, including an alternative to written exams ... ways to increase student responses to questions and discussion topics ... a student study clinic mini-course ... extra credit projects ... a way to involve students in correcting their own tests ... and more.

2. **Success-Directed Learning in the Classroom** shows how you can easily make your students accountable for their own learning and eliminate your role of villain in the grading process.

3. **General Classroom Management** provides solutions to a variety of management issues, such as laboratory safety, the student opposed to dissection, student lateness to class, and the chronic discipline problem, as well as innovative ways to handle such topics as keeping current in subject-matter content, parent-teacher conferences, preventing burnout, and more.

4. **An Inquiry Approach to Teaching** details a very effective approach that allows the students to participate as real scientists in a classroom atmosphere of inquiry learning as opposed to lab manual cookbook learning.

5. **Sponge Activities** gives you 100 reproducible activities you can use at the beginning of, during, or at the end of class periods. These are presented in a variety of formats and cover a wide range of biology topics, including the cell ... classification ... plants ... animals ... protists ... the microscope ... systems of the body ... anatomy ... physiology ... genetics ... and health.

343

And to help you quickly locate appropriate worksheets in Section 5, all 100 worksheets in the section are listed in alphabetical order in the Contents, from Algae (Worksheet 5-1) through Vitamins and Minerals (Worksheet 5-100).

For the beginning teacher new to the classroom situation as well as the more experienced teacher who may want a "new lease on teaching," ***Biology Teacher's Survival Guide*** is designed to bring fun, enjoyment, and profit to the teacher-student rapport that is called teaching.

About the Author

Michael F. Fleming, Ed.D., has taught biology, anatomy and physiology, microbiology, and behavioral science in the Council Rock School System of Newtown, Bucks County, Pennsylvania, for over 32 years. Dr. Fleming has presented papers at conventions of the National Association of Biology Teachers and has published articles in *The American Biology Teacher* and *Focus*. He is the author of *Life Science Labs Kit* (1985) and *Science Teacher's Instant Labs Kit* (1991), also published by The Center for Applied Research in Education.